Community Eco.

Community Ecology

R. J. Putman
Department of Biology, Southampton University, UK

CHAPMAN & HALL
London · Glasgow · New York · Tokyo · Melbourne · Madras

Published by Chapman & Hall, 2–6 Boundary Row, London SE1 8HN

Chapman & Hall, 2–6 Boundary Row, London SE1 8HN, UK

Blackie Academic & Professional, Wester Cleddens Road, Bishopbriggs, Glasgow G64 2NZ, UK

Chapman & Hall Inc., One Penn Plaza, 41st Floor, New York NY10119, USA

Chapman & Hall Japan, Thomson Publishing Japan, Hirakawacho Nemoto Building, 6F, 1–7–11 Hirakawa-cho, Chiyoda-ku, Tokyo 102, Japan

Chapman & Hall Australia, Thomas Nelson Australia, 102 Dodds Street, South Melbourne, Victoria 3205, Australia

Chapman & Hall India, R. Seshadri, 32 Second Main Road, CIT East, Madras 600 035, India

First edition 1994

© 1994 R. J. Putman

Phototypeset in Palatino by Intype, London

Printed in Great Britain by TJ Press Ltd, Padstow, Cornwall

ISBN 0 412 54490 3 (HB)   0 412 54500 4(PB)

Apart from any fair dealing for the purposes of research or private study, or criticism or review, as permitted under the UK Copyright Designs and Patents Act 1988, this publication may not be reproduced, stored or transmitted, in any form or by any means, without the prior permission in writing of the publishers, or in the case of reprographic reproduction only in accordance with the terms of the licences issued by the Copyright Licensing Agency in the UK, or in accordance with the terms of licences issued by the appropriate Reproduction Rights Organization outside the UK. Enquiries concerning reproduction outside the terms stated here should be sent to the publishers at the London address printed on this page.
    The publisher makes no representation, express or implied, with regard to the accuracy of the information contained in this book and cannot accept any legal responsibility or liability for any errors or omissions that may be made.

A catalogue record for this book is available from the British Library

Library of Congress Cataloging-in-Publication data available

∞ Printed on permanent acid-free text paper, manufactured in accordance with the proposed ANSI/NISO Z 39.48–199X and ANSI Z 39.48–1984

Contents

Preface   viii

1 Ecological communities – definitions and a search for pattern   **1**
   1.1 Biotic relationships between species   2
   1.2 The search for pattern   3
   1.3 The trophic structure of communities   5

2 Population interaction and the structure of communities   **14**
   2.1 The modelling of populations and their interactions   15
   2.2 Interference or exploitation: qualitatively different types of
          interaction   25
   2.3 Population effects of competitive or predatory interactions   27
   2.4 Empirical demonstrations of competitive exclusion or
          predator-prey extinctions   29
   2.5 Competitive coexistence: the role of spatial and temporal
          heterogeneity   31
   2.6 Searching for competition in real communities   33
   2.7 Indirect effects: diffuse competition, competitive mutualism
          and other more complex interactions   37

3 Food webs and connectance   **40**
   3.1 Important assumptions   41
   3.2 Trophodynamic implications of food web structure   42
   3.3 Internal structuring within food webs: problems of
          connectance   44
   3.4 Compartments in food webs   52
   3.5 Food web topology   52

4 Niche theory: niche packing and community structure — 60
4.1 The theory: a recapitulation — 61
4.2 The evidence? Niche shifts and character displacement — 64
4.3 Relationships in resource space: niche overlap, niche separation — 65
4.4 Limits to overlap — 71
4.5 Implications of niche dynamics for community structure: niche packing — 72
4.6 Initial doubts: problems of application and interpretation — 74
4.7 Additional problems: calculation and interpretation of multidimensional indices — 75

5 Guilds and guild structure — 80
5.1 Evidence for guild structure within communities — 83
5.2 Formal identification of guilds and their membership — 84

6 Species composition and the assembly of communities — 89
6.1 Short-listing candidates: limits to tolerance — 89
6.2 Attending for interview: dispersal — 91
6.3 Final election to membership — 93
6.4 Ecological succession — 94
6.5 Succession as a necessary statistical artefact — 98
6.6 Assembly rules — 100
6.7 Competition and the influence of invasion history on community assembly — 106

7 A question of equilibrium — 109
7.1 Determinism, stochasticism and chaos — 112

8 Species diversity — 115
8.1 Factors affecting species richness — 116
8.2 Are communities filled to capacity? Species richness below and beyond saturation — 119
8.3 Control of species richness: a unifying theory — 123
8.4 Species-abundance relationships — 126
8.5 The significance of patterns in relative abundance — 132

9 Stability — 135
9.1 Some definitions — 136
9.2 Stability of communities — 138
9.3 Effects of population interaction — 141
9.4 Diversity and stability — 145
9.5 Food web properties: the influence on stability of species richness, connectance and interaction strength — 147

Contents

9.6 Web topology and compartmentation	150
9.7 Succession and stability	151
9.8 Nutrient dynamics and energy flow	152
9.9 Environmental character	153
9.10 Causes of stability in ecological systems	154
References	158
Index	175

Preface

From the earliest development of ecology as a scientific discipline the sheer complexity of structure and interaction within communities seemed to preclude analysis of structure and function at the community level of organization. Early studies focused on the autecology of individual species, the population dynamics of single species populations and the effects on such dynamics of specific types of interaction (predation, competition, etc.) within simple interacting sets of two or three species; studies at the community level of organization were for the most part purely descriptive. Analytical studies perforce had to find some way of condensing the complexity or simplifying the structure; early attempts to identify patterns of organization or function at the community or ecosystem level considered the community simply as a system for energy transformations, collapsing the complex web of interaction into a simple hierarchy of trophic levels. Those who tried to retain in perspective the full complexity of interaction had to restrict analysis to some small subset of the entire community matrix – focusing attention on interactions between only a few members of the web: typically within a single trophic level, or single taxonomic assemblage.

Over the last two decades, however, there has been tremendous progress made in the study of ecology at the level of whole communities. The development of sophisticated modelling techniques capable of handling the complexity of interaction, together with increased rigour of analysis of field observation and recognition of the need for controlled experimentation have all led to tremendous advances in our understanding of the factors controlling the structure and composition of ecological communities. Yet progress has been so rapid that it is sometimes difficult for the interested bystander to keep pace.

In many areas of endeavour results of different approaches or from different methods of analysis seem contradictory; interpretation of much of the rest is highly controversial. Yet each researcher continues his own line of investigation and there have been few attempts to try to synthesize the major conclusions of all this work, to reconcile apparent confusion or contradiction. Much of the primary literature is in addition presented in a rather mathematical formulation – with important biological conclusions perhaps inaccessible to readers unfamiliar or uncomfortable with such treatment. In a comment on the completed manuscript of even this text one of my recent graduates wrote: 'Undergraduates have simple on/off switches. This is a protective device for use when any dangerous mathematics enters the field of vision. Concentration will, on seeing anything remotely resembling an equation, switch off until all signs of mathematical notation have gone and normal text has been restored.'

Over the years, in trying to offer undergraduate students some insight into this exciting area of ecology I have felt an increasing need for a simple text which attempts to draw together the complex and rather diffuse body of information now available to us on the organization and dynamics of ecological communities: a text which attempts a synthesis of the main issues, attempts to reconcile apparent conflict or contradiction and offers a simple overview of the major concepts and principles. It is in the hopes of offering such a synthesis to others that I have developed the present text. Throughout I have aimed to provide a brief overview rather than a comprehensive and exhaustive blockbuster: a roadmap of motorways and major trunk-routes, rather than a detailed gazetteer of all minor roads and byways; indeed this book is intended very much more as a framework for further reading than as a comprehensive treatment in its own right. It is aimed unashamedly at advanced level undergraduates. It is not a book for researchers; I myself am not an active contributor to developing theory in this particular field, but *am* accustomed to translating it to students. Equally, it is not directed to those who are just beginning the study of ecology. The text assumes throughout some familiarity with general concepts in ecology as might be provided by the majority of first and second year undergraduate courses or more general textbooks; each section in the book is self-contained and where prior knowledge is assumed, a brief recapitulation is offered of necessary background but such introduction is necessarily brief and offered as revision only.

My approach throughout has been a presentation of the fundamental concepts and principles underlying the structure and function of ecological **communities**. Even in the early stages of preparation of this book I received critical comment from referees that, for example, there was not enough coverage of 'microbial ecology' (sic) or 'plant ecology'. The rationalization of complexity in early community studies by taxonomic specialization has left ecology with an unfortunate legacy of taxonomic cliques – those who consider themselves plant ecologists, microbial ecologists, animal ecologists; others still, promoting such disciplines as marine ecology, insect ecology, bird ecology. Yet surely, in ecology in general – and especially in ecological study at the community level – preservation of such taxonomic distinction is unhelpful, even unproductive.

Most of the essential principles of population biology (to take just one example) apply equally well to insects, mammals, birds: is there any real need for each group of biologists to discover the wheel afresh? Likewise, those same principles apply (with necessary modification/translation) equally to plants, apply just as much in the marine or freshwater environment as in terrestrial ecosystems. Of course there are minor differences, due to real differences in organization of plants and animals or of environmental character between terrestrial and aquatic systems; but, rather than using those differences as an excuse for generation of a host of separate and specific models of population growth, they should surely be accommodated within a more general model applicable to all taxa, all environmental systems: a general model the more effective, the more robust because it does apply to all systems and taxa.

Certainly at the community level of analysis, such taxonomic freemasonry is necessarily myopic and frequently misleading. A community is generally defined as that part of any ecosystem embracing all living organisms and their interactions; by such definition, there is no such thing as a plant community, or an insect community. Further, focus at such a restricted taxonomic level will certainly underestimate, may miss altogether, important interactions with species beyond that limited taxonomic assemblage under study, which are nonetheless critical in influencing observed patterns of organization and function – may in the extreme recognize the patterns yet wrongly attribute the cause to interactions among the limited subset of organisms considered. Throughout this book, therefore, I have emphasized a *biological* perspective of the community as a completely integrated structure of plants, animals and microorganisms, and have attempted to explore patterns of organization, dynamics and function within that entire community. I can only hope such an approach pleases more people than it offends.

Because my intention is to offer an overview – a synthesis of the major issues and developments in community ecology over the past few years – I have deliberately kept the text brief and have attempted to keep it readable. Mathematical analyses are now such a part of research in community ecology that it would be impossible and inappropriate to strip this book altogether of all such material. However, they are introduced only where essential for understanding – and I have done my best on each occasion to 'translate' such formulations into English explaining what *biologically* they are attempting to represent. For helping me keep to these targets and ensuring my treatment is not shot through with too many inaccuracies of understanding or interpretation I am extremely grateful to the many people who have read chapters or complete drafts of the manuscript in various stages of its preparation. I would in particular like to thank Michael Fenner, Colin McHenry, Eric Pianka, Stuart Pimm and Tom Sherratt, for their comments and encouragement; any inadequacies that remain after their detailed criticisms remain entirely my responsibility!

I would also thank the following people for permission to reproduce in the text figures or tables copied directly from their own published materials: Dr J. Adams and Blackwell Scientific Publications (Figure 5.4); Drs S. J. Hall and D. Raffaelli and Blackwells (Table 3.1); Professor H. S. Horn and Blackwells (Tables 6.1, 6.2); Professor J. H. Lawton, Dr. P. Warren and the Elsevier Science Publishing Group (Figure 3.6 and Table 3.2); Professor R. M. May and Blackwell Scientific Publications (Figure 2.6); Dr M. Moulton, Professor S. L. Pimm and HarperCollins (Figure 6.5); Professor E. R. Pianka (with Blackwell Scientific Publications, Figure 4.4; with the Ecological Society of America, Figure 8.1); Professor D. Tilman and Blackwells (Tables 2.3, 2.4). Other illustrations were drawn specifically for this text but were based upon data published by others. I am most grateful for the very warm support and encouragement received from Dr P. A. Haefner (Figure 6.2), Professor J. H. Lawton (Figure 5.1), Professor P. B. Sale (Table 2.2), Dr W. P. Sousa (Table 8.1), Dr P. Stary (Figure 3.1) and Professor G. Sugihara (Figures 3.2, 3.4 and 3.5).

Rory Putman
Southampton, 1993

1
Ecological communities – definitions and a search for pattern

In an ecological context it is almost a gratuitous reaffirmation of faith to note that any individual organism or single-species population of organisms exists not in isolation, somehow separable from the complexity of interactions around it, but as an integrated component within a complex ecological 'whole' – one part only of an intricate mechanism of interdependent, individually moving parts. In effect, each organism or population is inextricably bound into a complex system of interdependence, influenced in its dynamics only partly through immediate interaction with the various abiotic features of its environment; affected also by the dynamics and activities of a multitude of other organisms around it – impinging upon their performance in its turn.

The precise pattern of resource use expressed by any organism, equilibrium population size, schedules of mortality and rates of growth, vary with context; all are determined or constrained by the specific characteristics of the particular environment in which it occurs. While, clearly, many of these constraints are imposed by physico-chemical features of that environmental context with a direct impact upon physiological efficiency or fitness, the other organisms which share that abiotic 'theatre' are equally a part of the total environment to which it must respond. More formally, each individual organism or population is seen within its context as part of a complete **ecosystem**. We may define such an ecosystem as a set of interacting species together with their physical environment: the smallest self-contained ecological unit of function. While considerable ecological analysis has been attempted at the level of the complete ecosystem, many studies restrict consideration to the living elements of that complex, generally considered as the ecological **community** of organisms: the assemblage of interacting species and the various interrelationships which bind them.

There are of course many definitions of such communities in the literature. Thus many phytosociologists have defined as 'communities' certain predictable associations of species: repeatedly encountered species-sets which were found commonly to co-occur. Such phytosociological assemblies by implication also embrace only the plant species contained within a given system; other authors too have used the term 'community' within a restricted taxonomic context, referring to 'insect communities', 'bird communities', 'animal communities'. Yet in our analyses the very essence of a community is that it should embrace the entire living 'fraction' of a given system, irrespective of taxonomic status, and that it is a functional unit, whose members interact.

Restricting consideration to passive species assemblies may include together species which merely co-occur. By the same token, restriction of focus to a limited taxonomic group may exclude from analysis extremely significant interactions, or may obscure indirect links critically important in determining the structure of a species assemblage, merely because those links involve members of a different taxon. Dispersion of an insectivorous or nectarivorous bird fauna across some resource continuum may contain inexplicable discontinuities in distribution, for example, merely because the observer has ignored bats or nectar-feeding rodents; regular co-occurrence of certain plant species within a sward might lead to misleading conclusions about mutualistic association were one to overlook the facilitating effects of a herbivore foraging selectively on the stronger competitor.

In our analyses then, we will adopt a more comprehensive definition of the community as the all-inclusive entirety of the biotic elements of an ecosystem: an interactive assemblage of species occurring together within a particular geographical area, a set of species whose ecological function and dynamics are in some way interdependent. This book will focus essentially on the structure and dynamics of communities so defined, the nature of the various interrelationships linking together the individual species, and the implications of the type, strength and direction of such interactions for each component species population and for the structure and function of the community as a whole.

1.1. BIOTIC RELATIONSHIPS BETWEEN SPECIES

The biotic relationships which may link together organisms within a community are many and varied; many of the linkages are also extremely subtle or ephemeral. Population and community ecologists only too commonly concentrate on competitive interaction or feeding relationships between species as the most significant associations linking community members, but we should not ignore the many other types of interactions which may occur, many of which, far from being trivial linkages, are critically important to the populations concerned. An animal or plant may rely on another for its habitat requirements; plants may rely upon animals for pollination or for dispersal of propagules. Plants may compete with other plants for light, space and nutrients; animals compete with each other for nest sites, shelter or food – as intimated, competition for resources offers a major area of interaction within the community. But we should not overlook the converse: the various cases of facultative or obligate mutualism where one species' performance is enhanced by, or even dependent upon, the presence of a symbiotic partner.

Feeding relationships, widely regarded as the primary medium of interaction between the various populations of a community, themselves embrace a variety of different interactions. Such relationships might be commensal, parasitic or holozoic; each will have a different impact on the partners involved in the interaction. Because such heterotrophic feeding relationships are in general readily apparent within a community, and so obvious in their effects upon the predator and prey, there has been perhaps too strong a presumption of the overriding importance of such interactions in structuring the relationships between the members of the community. A recurring theme of this book will be that we must not overlook the, often significant, impact of links within the community based upon competition, habitat requirements or phoresy.

1.2 THE SEARCH FOR PATTERN

The ecological community may then be characterized not simply as a flat species mosaic of co-occurring species but as a dynamic interactive system of interdependent populations. The parameters which define it are its **composition**: the species present and their relative abundances; **the nature and form of the relationships** between those species (the direction, relative strength and impact of those relationships); and its **dynamics**: its flux in both space and time.

While there should perhaps be no *a priori* presumption of any level of organization in nature above that of the individual (Roughgarden and Diamond, 1986), numerous authors have nonetheless recognized what appear to be certain regularities of pattern within such communities, certain constraints on structure and organization. In their excellent introduction to communities and the basic questions of community ecology, Roughgarden and Diamond (1986) review these apparent patterns of structure as: limited membership; restricted patterns of relative abundance among component species; correlations between body size and abundance; overdispersion of co-occurring species in resource space; patterns in food webs, limiting the total number of individual interactions in which any single species may be involved overall, limiting the number of links in any individual food 'chain', etc. (see also reviews by May, 1984, 1986a). If such apparent patterns, glimpsed as through the proverbial glass darkly, should prove substantive, they surely reflect real 'rules' of structure, limiting constraints on the organization and function of communities. It is that search for pattern and the explanation of perceived patterns that is in effect the essence of community ecology.

Many of the more degraded definitions of community which have been adopted over the years – as simple patterns of species co-occurrence, or single-taxon assemblages – have perhaps arisen from the sheer difficulty of carrying out studies that attempt to embrace the whole. The tremendous complexity of interrelationship within the community defies any attempt at definition or analysis. While it may prove possible to offer a detailed description of the composition and operation of a well-studied individual system – as, for example, the classic account of the ecological community of Wytham Woods offered by Elton in 1966, the very intricacy of such description of one single system itself confounds attempts to draw out any less specific principles of organization or function applicable to communities in general.

Historically this very complexity of structure forced those interested in ecological processes at the community level of organization into one or other of two very different approaches. Many adopted what might loosely be termed a reductionist approach, focusing attention on the dynamics of individual single-species populations. Such an approach considered intrinsic factors affecting population size, rates of growth, and periods of increase and decrease, later extending this perspective to consider the effects of the various relationships in which the focal population might be engaged with other populations around it. From such analysis of the dynamics of individual populations, and the influence upon those dynamics of interactions with other principal species (through predation, competition, etc.) one may attempt to reassemble a picture of the community and its operation as a jigsaw of component interacting parts. Of necessity analysis is restricted to the major species and the key interactions; the resulting picture of community dynamics can at best be a cartoon (although I use the word in the sense of Leonardo's cartoon of the Virgin, rather than Hanna-Barbara's Tom and Jerry!). The alternative, holistic approach attempts to study the community in its entirety, but of necessity must impose some simplification on the system, must attempt by some device to rationalize the picture in some degree. Such simplification may permit us to recognize certain similarities (or differences) of design and function of different communities and thus appreciate common denominators of structure or organization; but we should not forget that our conclusions cannot extend beyond the limits imposed by the level of analysis we adopt, the method of rationalization of the community's initial complexity that we have employed. 'Choice of the appropriate "macro-descriptors" or aggregate variables may be vital to progress; community ecology depends upon, *yet is simultaneously constrained by* the identification of such conceptual building blocks' (Pianka, 1980; my italics).

Despite these limitations, some of the early analyses of community structure did reveal some important underlying principles of organization. The bulk of this material receives exhaustive treatment in the majority of introductory ecology texts and will not be covered in detail here, but any study of communities would be sadly incomplete if it did not at least review some of the major conclusions of this early work.

1.3 THE TROPHIC STRUCTURE OF COMMUNITIES

In their search for a method of rationalization of complexity that would render their structure susceptible to analysis, many early authors (as Elton, 1927; Lindeman, 1942; Odum, 1957) adopted a trophodynamic approach to the study of communities and ecosystems. Communities were defined in terms of the feeding relationships between component species, and community members were divided up in terms of their trophic role. As we have already noted, such rationalization purely in terms of feeding relationships will leave not a few members of the community – and numerous additional interactions – unaccounted for. Nonetheless it is in essence possible to consider the community as a system for exchange of energy and organic matter between three basic types of organism: producers, consumers and decomposers. **Producers** are defined in such a scheme as that trophic level of autotrophic organisms within the community capable of synthesizing organic compounds from inorganic precursors and thus 'producing' food for the community. The community's **consumer** levels comprise those heterotrophic members of the community that feed upon the producers or upon each other. Within the community they may be separated out into **primary consumers** – those that feed directly upon the producer level; **secondary consumers** – those that feed upon the primary consumers, and so on. (In a terrestrial photosynthetic-based community, these different levels equate to the plants themselves (the producers), herbivores (primary consumers), small carnivores (secondary consumers), large carnivores, etc.) A final distinct group of organisms within the community may be recognized among the **decomposers**, saprophytic or saprozoic organisms which, as their name suggests, feed upon dead and decaying materials derived from the other two groups.

Organic matter, synthesized by the producers, was seen to be eaten by consumers at a series of levels; with the aid of the decomposers, all the organic materials incorporated into the bodies of the consumers (and indeed, unconsumed producers) are eventually broken down again and

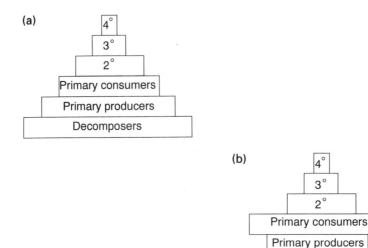

Figure 1.1 Typical pyramids of biomass derived for (a) a terrestrial green-plant-based community; (b) a freshwater detritus-based system.

returned to the producers for reuse. While available matter cycles within the system with no overall loss or gain, energy passes vertically through the trophic array in linear fashion: fixed by the producers in the chemical bonds of complex organic molecules, dissipated through the subsequent respiratory activity of the producers themselves, consumers and decomposers.

Of course such rationalization is oversimplistic. As we have already stressed, in focusing on feeding relationships – and only heterotrophic feeding relationships at that – it takes account of only one of the many types of interaction which occur within a community. Further, not all organisms fit tightly within a single category: a number may feed both as herbivores and carnivores, for example, or act as both consumers and decomposers. Nonetheless, despite such limitations, such rationalization does allow us to draw a number of important general conclusions about the structure and operation of such 'assemblies'. For instance, on the producer-consumer side of the system we find that, as first noted by Elton in 1927, organisms are structured into clear pyramids of numbers and biomass (Figure 1.1)

Since not all organisms in any given trophic level will be consumed by those of the trophic level above (indeed a large proportion may escape consumption); since in addition, not all material consumed may be assimilated and retained by the consumer (a proportion being lost in faeces and urine); and since of the fraction that is assimilated some must be devoted to maintenance and respiration rather than growth or production, there is a gross attenuation of available organic matter as one ascends the trophic scale, with the inevitable consequence that the total biomass of organisms at each trophic level will be substantially lower than that of the level immediately below. The gradual 'energetic attrition' in such feeding interactions, so that the energy retained by each consumer level is necessarily less than that available to it in the trophic level below, has also been invoked (Hutchinson, 1959; Slobodkin, 1961) to explain the further observation that there seems to be a limit to the number of possible links in any food chain and hence the number of trophic levels within the system. It is argued that the available food decreases so rapidly in translation that ultimately insufficient must remain in the system to support a further trophic level above. Although, as we shall see (Chapter 3), alternative explanations may be offered for this observation, the fact remains that in most terrestrial communities the number of trophic levels rarely exceeds four or five: producers and three or four levels of consumers.

Differences in structure are often more revealing than similarities. Perhaps more importantly, communities differ in the number of trophic levels they do contain, and in the relative importance of each trophic class within the community as a whole. Such differences can be clearly related to differences in 'economy' and mode of operation of these systems. Thus the stylised pyramid of Figure 1.1a is typical perhaps for terrestrial green-plant-based communities. By contrast, aquatic plant-based systems usually show a marked reduction in the consumer element, while certain other aquatic systems, which are not in themselves self-contained but rely on continuous input of material from outside in the form of detritus (fast-flowing streams; the lower, hypolimnion layer of deep lakes) show a totally different balance of trophic elements, with a vastly reduced producer level and with consumer levels supported largely by a dominant, detritus-based decomposer layer (Figure 1.1b). Such differences in trophic emphasis reflect different methods of operation in these differing communities and can be used as the basis for understanding their function. But the fact that communities of different 'function' show differences in structure also implies that the trophic organization of a community shows rather precise adaptation to that community's function. Such adaptation is persuasively argued in the now classic analysis of Heatwole and Levins (1972) (following Simberloff and Wilson, 1969), of trophic structure within the assembly of arthropods associated with tidal mangrove systems.

Simberloff and Wilson eliminated the fauna from several very small mangrove islands in the Florida Keys and then monitored their recolonization by terrestrial arthropods. In all cases the total number of species on an island returned to around its original value, although the species constituting that total were usually markedly different. Heatwole and Levins re-analysed these data in terms of trophic organization, listing for each island before defaunation and after recolonization the number of species in each of a number of (rather idiosyncratic) trophic categories: herbivores, scavengers, detritus feeders, wood-borers, ants, predators and parasites (Table 1.1). Examination of this table suggest a striking constancy of form. The distribution of species between different functional categories before defaunation is remarkably similar on each of the six islands recorded; similarly, the trophic structure on any island following recolonization approximates remarkably closely to the structure apparent before defaunation.

Such striking similarities in the trophic organization both between islands and within an island before and after recolonization are strongly suggestive of some 'ideal' optimal structure for operating as a mangrove community; despite their intuitive appeal however such arguments must be viewed with extreme caution. While the data marshalled in Table 1.1 seem to tell a most convincing story, more rigorous analysis has suggested that the trophic patterns observed are not significantly different from those expected to result from purely random colonization (Simberloff, 1976). While this does not mean that the structure observed is not adaptive in some way, it must at best leave the case for such adaptation unproven.

All such analysis is in any event essentially 'static': analysis of the components of structure of a system which by its very nature only has meaning in action, in operation. Some illustration of the importance of studying the community in operation in this way is offered by the observation of DeAngelis et al. (1978) and DeAngelis (1980) that community structures that would be assessed inherently unstable when viewed as static assemblies become progressively more stable as the turnover of energy within the system increases; in other words inherently unstable structures may in practice prove stable if energy flows through the system are high enough.

Four aspects of resource flow through the community may be considered here: amounts of energy 'handled' by the system (both amounts of material generated and amounts dissipated); efficiency of energy transfer between trophic levels; rate or speed of energy flow; and the nature of associated nutrient cycles (open or closed; sedimentary or non-sedimentary).

One of the major considerations here, as we have already hinted (page 7), is that of the efficiency with which energy and organic matter are transformed within the ecological system. As we have already noted, not all the energy an organism takes in in its food is retained as potential energy in growth; much is lost in assimilation and of that assimilated

Table 1.1 Evidence for apparent constancy of trophic structure in relation to function

Island	H	C	P	S	D	W	A	?	Total
E1	9 (7)	2 (1)	2 (1)	1 (0)	3 (2)	0 (0)	3 (0)	0 (0)	20 (11)
E2	11 (15)	9 (4)	3 (0)	2 (2)	2 (1)	2 (2)	7 (4)	0 (1)	36 (29)
E3	7 (10)	3 (4)	2 (2)	1 (2)	3 (2)	2 (0)	5 (6)	0 (0)	23 (26)
ST2	7 (6)	5 (4)	2 (1)	1 (1)	2 (1)	1 (0)	6 (5)	1 (0)	25 (18)
E7	9 (10)	4 (8)	1 (2)	1 (0)	2 (1)	1 (2)	5 (3)	0 (1)	23 (27)
E9	12 (7)	13 (10)	2 (3)	1 (0)	1 (1)	2 (2)	6 (5)	0 (1)	37 (29)
Total	55 (55)	36 (31)	12 (9)	7 (5)	13 (8)	8 (6)	32 (23)	1 (3)	164 (140)

The table shows the number of organisms in different trophic classes before defaunation and after recolonization of a series of island mangrove communities in Florida Keys; islands are labelled in the original notation of Simberloff and Wilson (1969). Trophic categories are the somewhat idiosyncratic classes adopted by Heatwole and Levins (1972) as: herbivore (H); predator (C); parasite (P); scavenger (S); detritivore (D); woodborer (W); and ants (A). For each, the first figure is the number of species before defaunation and the figure in brackets is the number after recolonization. Comparisons of the trophic structure of different island communities before defaunation and after recolonization suggest a striking constancy of trophic organization. (After Heatwole and Levins, 1972.)

energy more is expended in respiration. Nor is the entire production of one trophic level passed to the trophic level above; much is unavailable to the consumers and many individuals will die unconsumed. If we look at the efficiency of energy transfer between trophic levels as

$$\frac{\text{Energy consumed by trophic level n}}{\text{Energy consumed by trophic level n-1}}$$

we will always find its value to be considerably less than one. Indeed in terms of percentage, this **gross ecological efficiency** is generally of the order of between 7–14%. It has even been suggested that as a general rule of thumb one may presume gross ecological efficiency to be a constant at 10%. Yet such overgeneralization is as obscuring as it is misleading; it is the differences in efficiency which are the more interesting. Efficiency of energy transfer may differ between different pairs of trophic levels within one community, or between equivalent pairs of trophic levels of different communities. Such differences offer an insight into the different mechanics and efficiency of operation of different trophic classes or differences in operation between communities of subtly different style (Table 1.2).

Table 1.2 Efficiency of energy transfer (per cent) at various trophic levels in three aquatic ecosystems

Trophic level	Cedar Bog Lake, Minnesota	Lake Mendota, Wisconsin	Silver Springs, Florida
Photosynthetic plants (producers)	0.1	0.4	1.2
Herbivores (primary consumers)	13.3	8.7	16
Small carnivores (secondary consumers)	22.3	5.5	11
Large carnivores (tertiary consumers)	not present	13.0	6

Figures are shown for the Gross Ecological Efficiency (see text) at different trophic levels in three aquatic ecosystems (after Odum 1959). Comparison is deliberately restricted to systems of relatively similar structure and function in order to minimize obvious differences between systems of grossly different economy; despite this, considerable variation is apparent in efficiency of energy transfer at different trophic levels within any one system and for any given trophic level in the three different communities.

In the same way, the quantities of energy 'processed' by a community and the rates of energy flow may themselves reflect particular characteristics of organization within that community, or may in turn have implications for its structure and stability. We have already noted the importance of rates of resource flux through the system in permitting the persistence of inherently unstable structures; in the same way the amount of energy passed through the system at any one time may reflect or influence community organization. Communities, as we have already suggested, may be dependent on external inputs of energy, such as detritus, or may be self-contained, dependent on internal energy production. The total amount of energy captured within a community by the producer level is usually referred to as the **gross primary production**; that available to other elements of the community after respiratory expenditure by the producers (GPP − Respiratory cost) is referred to as **net primary production** (NPP).

That GPP − and thus the amount of energy passing through the whole community − can vary enormously between different systems is graphically illustrated by Figure 1.2, while Odum has suggested that such variation in the type, quantity and quality of energy flows through the community may have very profound implications for its whole structure and mode of function. In 1974 he wrote

> The species matrix [of the community] adapts to the strength and variety of energy input and the resource flows coupled with it.... Furthermore, the quality of the energy in terms of utility and low entropy is as important as the quantity. When one or a few sources of high utility energy ... are available ... low diversity has advantages; a concentrated and specialised structure is more efficient in exploiting the bonanza than is a dispersed structure. Where energy is limiting or of low utility, then a higher diversity appears to be optimum for [the operation of] a steady-state system.

These links between energy flows and diversity (the implications for species diversity and connectivity within a community web, of low and predictable internal production of low utility energy or high, but unpredictable inputs of high utility energy derived from external sources) will be examined further in later sections.

However informative such a trophodynamic level of rationalization may have proved in understanding the community as a level of organization, it tells us little about the detailed structuring of such communities. As we noted on p. 5, community ecology depends upon, but is simultaneously constrained by the identification of the appropriate macrodescriptors for rationalization. Treatment of the community as a series of trophic classes obscures consideration of its organization within each trophic level: the relative numbers and relative importance of individual species to the trophic class as a whole. In addition, while in the real world relationships between trophic levels are made up of a myriad of separate interactions between individual pairs of organisms in the different levels, this complexity of interaction is crushed into some unitary

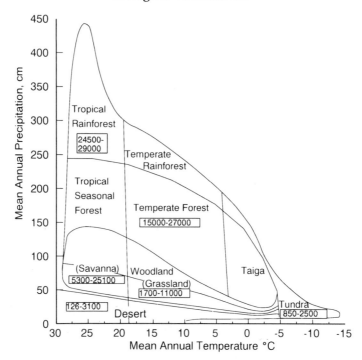

Figure 1.2 Gross Primary Productivity of terrestrial ecosystems. The major vegetational systems of the world may be relatively simply characterized with reference to two main environmental variables: mean annual temperature and mean annual precipitation (after Whittaker, 1970); superimposed on the diagram are representative values for the gross primary production (kJm^{-2}yr^{-1}) for a number of such systems (values derived from McNaughton *et al.* (1989)).

relation between one trophic block and another. Yet again, as we have already noted, many organisms are in any case not restricted to a single trophic level. With the advent of more sophisticated artificial intelligence, it has been possible to gain some insight into the details of this more subtle structuring (while still retaining some measure of simplicity) in the analysis of food web design in different communities. Such studies permit examination of community structure at a slightly more biologically meaningful level of complexity than does pure trophic level analysis – but we should remember that consideration is still restricted to only one of the many forms of interaction between the various organisms of the community.

In essence the approach to such studies of food web design is to generate random assemblages of populations in computer models and then look for common characteristics among the subset of such randomly assembled model webs that show some measure of stability, finally comparing these shared characteristics of stable models with the properties of real-life communities. An overview of the approach is presented by Paine (1980) and a summary of the main conclusions of such analyses will be presented in Chapter 3.

Such studies of food web design are based upon a principle of construction of model communities through the step-by-step assembly of a set of interacting species populations. In effect it exemplifies the rationale of the whole reductionist approach to community analysis: piecing back together into a composite whole, the separate jigsaw pieces of a community matrix whose dynamics and inter-relationships have been studied in isolation. In this case community matrices may be formed by modelling the population dynamics of a single-species population – subject only to intrinsic rates of increase and schedules of mortality, resource limitations and the processes of intraspecific competition. Simple two-species models may be presented where the dynamics of two populations interacting through competition or predation may be addressed. These same models may be extended to accommodate a third interactor, and gradually we may develop a more complex matrix of interaction. But, as noted above (page 4) such models are inevitably restricted to a simple 'cartoon' of community dynamics – and such is the process of assembly that at each stage any errors of precision or incorrect assumptions are compounded as more and more species are added to the matrix.

Finally, we should note that all that is modelled in such community matrices are the actual population dynamics of their component species; nothing is resolved regarding relationships or interactions in resource use, of partitioning of available resources between the species, or adjustment of patterns of resource utilization to take account of the resource use functions of other species. These types of resource relationships have, however, been explored in other reductionist models in an analysis of the resource use patterns of individuals and populations, niche relationships of co-occurring species, limits to niche overlap, and limits to packing of niches in resource space (e.g. MacArthur and Levins, 1967; May and MacArthur, 1972; Roughgarden, 1974; Roughgarden and Feldman, 1975; Pianka, 1975; Rappoldt and Hogeweg, 1980). Such models, the insights they offer on rules of structure, and their limitations, will be discussed in detail in Chapter 4.

2

Population interaction and the structure of communities

The effects of different types of interaction amongst the community's component populations (section 1.2) may be modelled in terms of their consequences for population growth and dynamics – allowing us to consider in more detail the precise implications of such interaction for the populations concerned. The models used to describe the dynamics of such populations fall in effect into two distinct 'families', based on differential equations (modelling populations with continuous growth) or difference equations (used to model populations with more discrete patterns of growth and non-overlapping generations). A whole body of research effort has been devoted to such analysis of the dynamics of single-species populations and the consequences of interactions between species. The broader implications of this line of investigation will be explored more fully in later chapters; in preparation for such analysis, it may help to offer here a brief introduction to the study of population dynamics and the major models used.

Population ecology is however a vast and complex field of study in its own right; my aim here is simply to review the barest essentials of population dynamics and models of population growth, sufficient merely to allow us to develop some understanding of how interactions between populations may be caricatured in appropriate mathematics to facilitate exploration of the dynamics of simple interactions. For a more definitive treatment the reader is referred to more specific texts such as Begon and Mortimer (1992), or the population ecology sections of more general introductory texts.

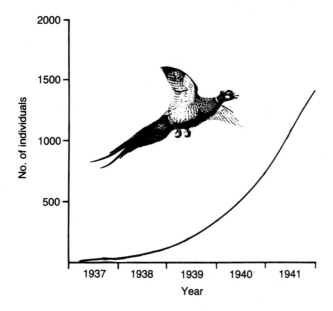

Figure 2.1 Exponential growth of pheasant (*Phasianus colchicus*) populations on Protection Island, Washington. Smoothed curve connects average numbers at the end of each year. (After Einarsen, 1945)

2.1 THE MODELLING OF POPULATIONS AND THEIR INTERACTIONS

In an unlimited environment, population growth will follow a simple geometric progression as the full natural rate of increase of that population is expressed; if in one generation some imaginary population doubles in size, in the next generation that double population will double again, and so on. This sort of increase leads to a pattern of population growth such as that shown in Figure 2.1. The growth of such a population can be very simply described by a differential equation:

$$\frac{dN}{dt} = r_m N$$

[Equation 1]

where dN/dt, the rate of change of numbers over time, equals the product of the natural rate of increase, r_m, and population size at any instant, N.

Such growth clearly cannot be maintained indefinitely; as resources become limiting, intraspecific competition has a dampening or depressive

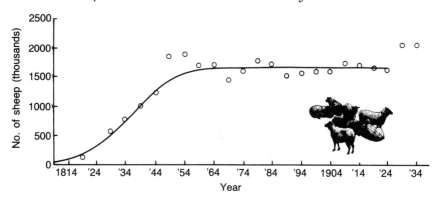

Figure 2.2 Typical sigmoid curve of population growth with restricted resources. The graph plots the development of sheep populations in Tasmania. (From Davidson, 1938, after Odum, 1959.)

effect on such free growth. We may incorporate a second element into our equation to represent this dampening effect of intraspecific competition, as

$$\frac{dN}{dt} = r_m N \left[1 - \frac{N}{K}\right] \quad \text{[Equation 2]}$$

when N is again instantaneous population size, r_m is intrinsic rate of increase and K is a measure of the total population size that the environment can support at balance: the **carrying-capacity** of that environment. This new equation – developed by Verhulst (1838) and later and independently by Pearl (1925) – models a population growth curve of the familiar sigmoid form of Figure 2.2, where after an initial period of rapid increase, growth rate declines and the population levels to a plateau: a situation far more commonly experienced in nature.

(Note that it is easy to understand biologically what the model is doing mathematically, by considering what will happen to our theoretical population as the ratio of N to K changes. Where $N < K$, and thus population size is below its carrying capacity, N/K is less than one, the function $(1 - N/K)$ is thus greater than zero and the population will increase in size. Indeed when N is very small, N/K tends to zero and the population will express its full intrinsic rate of increase, with no dampening effect of intraspecific competition experienced. Where $N = K$ the population will remain stable $(1 - N/K = 0$; thus rate of growth is zero). Where $N > K$, the population will decline.)

Into this simple model it is possible to incorporate, in much the same way, a further dampening function to represent the effects of, for example, interspecific competition. In its simplest form this term is calculated to include an effect from the size of the population of the competing species (N_2) and a competition coefficient (α), representing the strength of interaction by altering the extent to which growth rate of species 1 is influenced by the presence of a given number of species 2. This competition coefficient is likely to depend, *inter alia*, on factors such as the extent to which resource use by species 2 overlaps with that of species 1.

Our equation for population growth of species 1 now becomes:

$$\frac{dN_1}{dt} = r_1 N_1 \left[1 - \frac{N_1}{K_1}\right] - r_1 N_1 \left[\alpha_{1,2} \frac{N_2}{K_1}\right] \quad \text{[Equation 3]}$$

or:

$$\frac{dN_1}{dt} = r_1 N_1 \left[1 - \frac{N_1}{K_1} - \alpha_{1,2} \frac{N_2}{K_1}\right] \quad \text{[Equation 4]}$$

K_1 appears in the denominator throughout since we are interested in the effects of both intraspecific and interspecific competition on resources used by species 1.

The growth of species 2 can equally be described as:

$$\frac{dN_2}{dt} = r_2 N_2 \left[1 - \frac{N_2}{K_2} - \alpha_{2,1} \frac{N_1}{K_2}\right] \quad \text{[Equation 5]}$$

Clearly, these very simple equations can be extended to accommodate interaction with more than one competing species. Additional elements ($-\alpha_{n,m} N/K$) may be added to represent the additional effects of competition experienced from a third, fourth or fifth competitor. In addition, these same basic equations may be modified to include the effects of other types of interaction on population growth such as predation, parasitism or mutualism. Positive effects on population growth of mutualistic partners may be represented by altering the signs of the interaction coefficients ($\alpha_{1,2}$, $\alpha_{2,1}$ elements of equations 4 and 5) to positive rather than negative. Parasitic interactions can be represented by altering the sign to + in one case (representing the parasite's population) and − in the other (indicating the negative effect upon the host population's rate of growth) and equations for predator-prey interactions can be derived in a similar way. Indeed all the various relationships through which the various members of a community might interact, may be represented as ++ (mutually beneficial); +0 (beneficial to one participant, not affecting the other); and +− (beneficial to one, damaging to the other). Under such a scheme the various interactions discussed so far can be simplified and, more importantly, grouped into relationships of similar effect, as in Table 2.1 (see also Arthur and Mitchell, 1989). Such rationalization is particularly useful, for, as well as summarizing the various possible interactions and classifying together relationships of like type, it also carries implicit within it an indication of the effect that any particular relationship may be expected to have upon the 'performance' of the participants: increasing (+) or decreasing (−) population growth, allowing us to incorporate it easily into the models of population growth we have developed above.

Table 2.1 Schematic representation of the various types of biotic relationships which may link together the different organisms of a community

Effect of relationship on participants	Examples of relationship
+0	Use by one organism of another as habitat Commensal feeding relationships
++	Pollen-feeding and thus pollination by animals, or fruit-feeding by animals resulting in seed dispersal; mutualistic feeding relationships (e.g. symbiotic algae in green hydra)
+−	Heterotrophic, and parasitic feeding relationships Secretion of allelopathic chemicals by plants, preventing others growing around them
−−	Competition
0−	Incidental damage: non-deliberate damage to individual organisms or their environment (e.g. trampling by animals causing damage to plant species)

Relationships are grouped as mutually beneficial (++); one sided (+−); mutually disadvantageous (−−), etc.

The analyses developed here upon a logistic model of population growth were first suggested by Lotka (1925) and Volterra (1926) on whose work these simple equations are based. It should be emphasized that it is possible to derive similar models based upon other mathematical premises. The Lotka–Volterra equations greatly oversimplify the processes of population growth and interaction and a number of alternative models, or refinements of the simple Lotka–Volterra models as presented here, have been proposed. However, the basic logic of these models is essentially the same, and a preponderance of community-level models based upon reassembly of a series of interactive populations depend upon simple Lotka–Volterra mathematics (Chapter 3).

Strictly, however, such mathematics is restricted to modelling the dynamics of populations exhibiting continuous growth; a series of equivalent models has been developed for populations with discontinuous growth based upon difference equations.

Modelling populations and their interactions

In such models time becomes a discrete variable and the general form of such equations is

N_{t+1} = function (N_t)

(Box 1)

> **BOX 1**
>
> Differential population models of the Lotka–Volterra form are strictly only applicable to populations exhibiting continuous growth; a series of equivalent models has been developed for populations with discontinuous growth based upon difference equations.
>
> In this type of model, equations for unrestricted population growth may be expressed as, for example:
>
> $$N_{t+1} = \lambda N_t \quad \text{[Equation 6]}$$
>
> where N_t or N_{t+1} represent the numbers of organisms in the population at times t and $t+1$ respectively and λ, the finite rate of increase, is the number of times the population multiplies itself each generation. $\ln \lambda = r$, the difference equation equivalent of the intrinsic rate of increase of a continuously growing population, and Equation 6 may be rearranged as
>
> $$N_{t+1} = N_t e^r, \text{ or } N_t \exp^r \quad \text{[Equation 7]}$$
>
> Into such an equation for unrestricted growth we can incorporate an element of negative feedback due to intraspecific competition in a very similar way to that applied for the Verhulst–Pearl logistic equation as:
>
> $$N_{t+1} = N_t \exp[r(1-\underline{\frac{N}{K}})] \quad \text{[Equation 8]}$$
>
> Like the logistic equations, too, these difference models can be extended to incorporate the effects on population growth of various types of interaction with other species. A simple model of predator–prey relationships, or parasite (P) – host (H) dynamics may be presented, (as Beddington, Free and Lawton's (1975) modification of the original model of Nicholson and Bailey, 1935) as
>
> $$H_{t+1} = H_t \exp\left[r\left(1 - \frac{H_t}{K}\right) - q P_t\right] \quad \text{[Equation 9]}$$
>
> $$P_{t+1} = H_t [1 - \exp(-qP_t)] \quad \text{[Equation 10]}$$
>
> Where K is the carrying capacity of the host (prey) population in the absence of parasitoids (predators) and q is some representation of the parasitoid's efficiency in discovering host individuals. For the Lotka–Volterra equations we developed models of interactive populations by reference to interspecific competition; the difference equa-

> tion equivalents of such competition equations may be represented in Hassell's general model (Hassell, 1975):
>
> $$N_{t+1} = \frac{r N_t}{(1 + aN_t)^b}$$
>
> [Equation 11]
>
> [where r is the net reproductive rate, $a = (r-1)/K$, the element required to incorporate some density dependent control of the net rate of increase due to intraspecific competition, and b is an exponent of the strength of interspecific competition measured as the additional reduction in population growth and estimated as the slope of the relationship $\log_{10}(1+aN_t)$ vs. $\log_{10}N_t$].
>
> These formulations may appear somewhat more complex than the relatively intuitive mathematics of the logistic, Lotka-Volterra equations. In practice the conclusions that may be drawn from such models for discontinuous population growth show many parallels with those derived from differential models.

Both types of model predict responses of the population in relation to changes in the numbers of individuals. While they are readily applied to most animal populations, their application to plant populations is not so straightforward, since for many plant species with a colonial or clonal growth form it is very hard to distinguish the individual. Further, plants may reproduce both by production and dispersal of seeds (new individuals) or by vegetative reproduction. In addition, plants often respond to density by a plastic reduction in individual size as well as by self thinning. These problems are discussed by Harper (1977, 1981) and Watkinson (1986). Specific models relevant to plant populations have since been developed by, for example, Watkinson (1980, 1986); these differ in detail rather than essentials from the models we have already introduced (Box 2).

To conclude this brief review we may note that all the models considered so far consider all members of the population to be equivalent – no account is taken of age-structure or possible age-specific differences in fecundity or survival. It is possible to model the dynamics of both animal and plant populations characterized by discontinuous growth dynamics and distinct age-classes or growth phases using matrix algebra. Models for animals have been developed by many authors following the original presentations of Leslie (1945, 1948); such models have been applied to plant populations by, among others, Werner and Caswell (1977), Harper and Bell (1979), and Silvertown (1982). Age-structured populations with continuous growth may also be modelled, following Gurney, Nisbet and Lawton (1983).

BOX 2

In general, seed plants show discrete patterns of population growth and are more accurately modelled by difference equations like those of Box 1. Watkinson (1980) proposes a model for the population growth of simple annuals as:

$$N_{t+1} = \frac{\lambda N_t}{(1 - cN_t)^d + m\lambda N_t}$$

[Equation 12]

where λ, as before, is defined as the finite rate of increase of the population – and is defined here as the mean number of seeds produced by plants in isolation and their probability of surviving density independent mortality. m^{-1} defines the maximum density of plants any given habitat can support, while c is the area required for each plant and d its efficiency of resource utilisation. (Notation here is changed from the original formulation by Watkinson (1980) to avoid confusion with equation 11, Box 1.) The equation is in essence a simple modification of our initial exponential growth curve (equation 7) by incorporation of two density dependent elements, a term (m) for self-thinning and a term $(1 + cN)^{-d}$ to describe reductions in plant size with increased density.

A separate equation,

$$w = w_m (1 + cN)^{-d}$$

[Equation 13]

describes the effective yield or biomass of those plants.

Competition between plant species may be represented from these equations (Watkinson 1981), as:

$$w_1 = w_{m1} [1 + c_1(N_1 + \alpha_{12}N_2)]^{-d_1}$$

[Equation 14]

and

$$w_2 = w_{m2} [1 + c_2(N_2 + \alpha_{2,1}N_1)]^{-d_2}$$

[Equation 15]

where $\alpha_{1,2}$ and $\alpha_{2,1}$ are our familiar competition coefficients, and competition is considered in terms of its effect on population biomass not individual number.

As modifications may be made to earlier models presented, to accommodate the additional effects of predatory interaction, these plant-population models may be extended in a similar way to model the effects of herbivory.

Graphical models of competitive interaction

Although resource competition theory is best expressed using mathematical models of this kind, its major features can be understood graphically. Tilman has developed a graphical model of competitive interaction based upon analysis of population growth isoclines (Tilman, 1980, 1982, 1986). Because they are not based on the specific assumptions of Lotka–Volterra or any other simulation models, Tilman's models of competition are in many ways more general in their application and interpretation; further, the models can consider competitive interaction on a number of independent resource axes simultaneously and are developed to consider both rates of resource consumption and resource renewal. It is beyond the scope or intention of this book to develop Tilman's arguments in any great detail and the models are explored more fully elsewhere (Tilman, 1982, 1986; general textbooks such as Begon, Harper and Townsend, 1986); we may however offer here a brief overview of the approach.

For each species in Tilman's models we may define a per capita rate of reproduction $(dN/dt)/N$ dependent on the availability of resources (for simplicity we will consider only two resource types for now, R_1 and R_2). For the species to persist in any given environment its reproductive rate must exceed the mortality rate, m. Thus in Figure 2.3, if species A lived in a habitat where it experienced the mortality rate indicated by m_A the rate of mortality would precisely balance the per capita rate of reproduction for levels of resource R^*_{A1} on Resource 1; R^*_{A2} on Resource 2. For resource levels which exceed this threshold, population density of species A would increase – but at the same time the rate of consumption of the resource would increase too, and resource availability per capita would thus decline. For resource values below R^* population density of species A would decrease, but by the same token, since resource consumption rates fall, availability of the resources will increase over time. In the absence of interspecific competition the population density of species A would equilibrate at the point where its rate of consumption of the most limiting resource equalled the rate of supply of that resource and where reproductive rate equals birth rate. However, when species A experiences competition for either one or both resources from a second species, consumption of resources by species B affects the supply of resources available to species A, and may push the availability of either resource below R^*_A (the equilibrium point for species A).

Consider species A of Figure 2.3. It requires R^*_{A1} (one unit, in our example) of resource 1 and R^*_{A2} (three units) of resource 2 in order to maintain a stable population. These requirements define the growth isocline presented in Figure 2.4a. For all environments where the availability of R_1 and R_2 exceed those lower limits, the population density will increase; for all environments where the availability of either R_1 or R_2

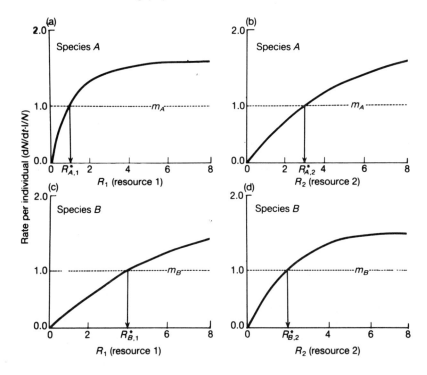

Figure 2.3 Tilman's graphical models of resource use competition. Each part of the figure represents the per capita rate of reproduction and per capita mortality rate of some hypothetical species (A and B), against the availability of some resource (R_1, R_2). For each species and each resource the level of resource availability at which mortality equals recruitment is indicated by the arrow pointing to equilibrium resource availability R*. The four R* values from this figure are used to construct the competition graphs of Figure 2.4

falls below the critical limits, population density will decline. The solid line superimposed upon Figure 2.4a represents the isocline for zero growth; for all environments that have availabilities of R_1 and R_2 that fall on this isocline the reproductive rate of species A will equal its mortality rate. (The right-angled form of the isocline indicates that species A is limited either by R_1 or R_2, which ever is most limited in supply in relation to its requirement.) Equilibrium will occur when resource consumption equals resource supply and reproduction equals mortality. Reproduction equals mortality for any point on the isocline; the actual point on that isocline which will represent equilibrium for any given environment depends on rates of resource consumption in relation to supply.

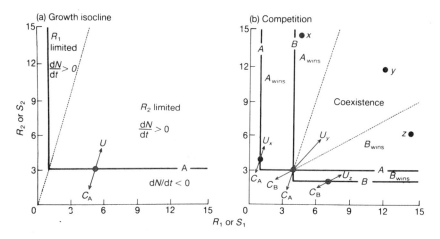

Figure 2.4 Resource use competition graphs for two hypothetical species. (a) The solid curve with the right angle corner is the resource dependent growth isocline of species A when utilising two resources in isolation, derived from resource use curves of Figure 2.3 a) and b). At all points on the isocline, reproductive rate equals mortality rate. Above the isocline (further away from the origin) it increases in abundance; below or to the left of the isocline its declines. The broken line from the origin through the corner of the isocline shows the levels of R_1 and R_2 where species A is equally limited by both resources. The dot on the isocline is the equilibrium point for species A where reproductive rate equals mortality rate *and* resource consumption rate (C) equals resource supply rate (U). (b) growth isoclines for species A and species B in competition. The positions of the isoclines for A and B (determined by their different R* values for each resource; Figure 2.3) define those conditions where species A increases while species B declines, where species B is dominant and where both may theoretically coexist. (After Tilman, 1982, 1986; see text for details.)

Let us now superimpose upon the figure an equivalent representation of population growth, and resource consumption, by species B (Figure 2.4b). Once again equilibrium resource values for R_{1B}, R_{2B} (three and two units respectively) are taken from the arbitrary example of Figure 2.3. The area outside the intersection of the isoclines for both species A and B is a zone where theoretically both species could coexist: reproductive rates exceed mortality rates for both species A and B. However, such potential coexistence is in some sense a trivial case; it equates to a situation where resources are in any case superabundant. If environments have a low supply rate of R_1 and high supply rate of R_2 both species will be limited by R_1. Since the equilibrium requirement $R*_1$ is lower for species A than for species B, species A will outcompete B: its population can continue to grow (and continue to consume resources, reducing still further their overall availability) at resource levels of R_1 where species B is in decline. Similarly, species B is a superior competitor for R_2 and will displace species A from environments in which both species are limited by R_2.

2.2 INTERFERENCE OR EXPLOITATION: QUALITATIVELY DIFFERENT TYPES OF INTERACTION

Most models of competition expressly consider the effects on population growth of any one species of the sequestration of some portion of its required resource base by another species. Explicitly, therefore, the majority of these models consider competition only from the viewpoint of an exploitative interaction: the effects upon the population dynamics of one species of a reduction in the level of available resources through consumption by some other species.

Yet, among animals certainly, and possibly among plants, competition may also involve an element of interference: physical interference in freedom of access to those resources. Individual animals may coincide in time and space in their attempts to exploit the same patch of resources, promoting some direct behavioural interaction. Plants, it is argued, may engage in similar interference interaction in the production of chemicals that are toxic to other species (**allelopathy**). There is no doubt that chemicals with such properties may be produced by some plants, but the role of such chemicals in competitive interactions is still debated (reviews by Harper, 1975; Crawley, 1986).

In effect, we may recognize within the context of competition two distinct types of interaction. Competition may be direct, through direct interactions of this sort between individuals seeking to use the same resource and thus interfering with each other's access to that resource (**interference competition**), or it may be indirect, with one individual affecting the other's free use of some resource through prior consumption (**exploitation competition**). Often a competitive interaction will contain elements of both interference and exploitation but it is important to distinguish the two types of interaction, for they have somewhat different effects. Indeed the form of interaction involved in any instance may itself reflect differences in response in relation to contests over qualitatively different classes of resource (page 7–8; see also Carothers and Jaksic, 1984). For interference interaction, competitors would usually have to coincide in their use of some divisible or indivisible resource. In exploitation interactions, the competitors do not necessarily ever have to meet and thus directly restrict each other's use of the resources contested; whether they occupy the resource simultaneously or sequentially, each nonetheless affects the other's subsequent use of any potentially exhaustible resource by reducing the total amount available.

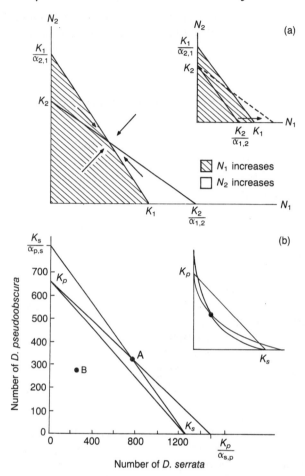

Figure 2.5 Interference and exploitation competition. (a) Graphical models for the outcome of competitive interactions similar to those of Tilman (1982, 1986) may be derived from the Lotka-Volterra competition equations. In effect they are a graphical representation of the solution of the logistic competition equations for all values of N_1 and N_2. (Thus for given K the value of N_1 is derived for all values of N_2 and vice versa.) The arrows superimposed in a) show zones where populations of N_1 and N_2 may increase or must decline; the point of intersection of the two isoclines represents the equilibrium point for coexistence and shows the expected population sizes of the two species based on exploitation interactions. (b) The same graphical models drawn to represent the outcome of a real competitive interaction between two species of *Drosophila*, *D. pseudo obscura* and *D. serrata* (Ayala 1970). Logistic equations modelling only exploitation interactions predict equilibrium population sizes of the two species in coexistence as at A. Actual population sizes recorded are shown by B. The inset shows the effect on population trajectories of the two species of incorporating an element of interference competition into the models with variable α.

As noted, the majority of population models describing competitive interactions, implicitly or declaredly focus upon implications of resource depletion through common exploitation and it is recognized that such models sometimes predict but poorly the outcome of competitive interactions in the real world. The population implications of interference interactions are bound to depend on the sizes of the two competing populations; this must mean that the interaction coefficients summarizing the strength of interaction (the α coefficients, for example, of the simple Lotka–Volterra equations 4 and 5) should be variable with respect to population size. In most formulations such coefficients are constant, and reflect merely the relative resource preemption abilities of the two competing species; implications of population number of the competitors (e.g. N_2/K_1) are restricted to consideration of the resultant level of resource depletion imposed by a given number of competitors with a fixed individual exploitation efficiency. Incorporation of a variable competition coefficient into population models of this sort – derived as some function of relative population size – restores some element of interference to the interaction and improves their predictive power (Figure 2.5).

2.3 POPULATION EFFECTS OF COMPETITIVE OR PREDATORY INTERACTIONS

Introduction of an element of interspecific competition or predation to the dynamics of a single-species population not only results in a reduction in the rate of growth of such a population, but, in general, is also accompanied by increasing instability of the population's dynamics by comparison with one where the dynamics are influenced only by intrinsic feedback from intraspecific competition. A whole range of single-species population growth patterns may be generated from the difference or differential models we have just introduced, merely by varying the values of r and the starting population (Figure 2.6). The population trajectories of such single-species models may show monotonic damping to some stable equilibrium point, damped oscillations around the same single equilibrium point, two- or four-point limit cycles, or chaotic behaviour, depending on the specific parameter values. In general, predatory interaction causes a shift in the dynamics of such single-species populations to a lower stability 'zone' (Figure 2.7). Competitive interactions are also generally considered destabilizing. In the limit, both types of interaction can cause extinction of the population of one or both species.

Introduction of such interaction into the community web may therefore have profound consequences; in response to the risk of predation or competition, actual or potential, organisms may adjust their patterns of resource use to minimize the frequency or level of interaction, may show markedly reduced abundance, or may in the limit be excluded from the

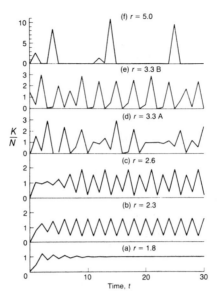

Figure 2.6 Dynamical Behaviour of Populations. Changes in the population density of organisms over time show a range of different dynamics: with damped oscillations around some single equilibrium density, two– or four– point limit cycles, or erratic and chaotic dynamics. The range of possible outcomes is here illustrated by analysis of dynamical behaviour of population density (N_t/K) modelled by differential or difference equations; the nature of the dynamics is influenced in such models by the precise value of r (the intrinsic rate of increase). (From May, 1976.)

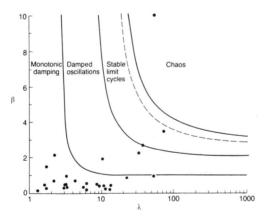

Figure 2.7 Population stability. The alternative outcomes for the dynamics of populations shown in Figure 2.6 are expressed here by derivation from difference equation models as zones on a two-way plot of β against λ (from the specific difference model of Hassell, Lawton and May, 1976). The solid curves separate the regions of monotonic damping, damped oscillations, stable limit cycles and chaotic behaviour; the broken line indicates where two-point cycles give way to four-point cycles. Filled circles represent the actual position of a number of field and laboratory insect populations

community altogether. Even if competitive or predatory interactions may relatively rarely lead to extinction, their immediate effects in suppressing population growth and potential must surely constitute a significant selection pressure on prey populations and on populations of both species engaged in actual or potential competitive interaction. Such selection pressures may lead to changes in ecology of resource use, as we have already suggested, both in the short-term (through behavioural modification of resource use patterns) and/or in the longer term, through evolutionary adaptation, if such shifts in resource use are maintained. Parasitism and herbivory still more rarely threaten the survival of the affected population, and indeed seldom result in the death of the individual host; yet both again will have a significant depressing effect upon performance and individual fitness and may result in evolutionary adaptation and counter-adaptation in those species involved in the interaction (e.g. Edwards and Wratten, 1980; Crawley, 1983; Futuyma and Slatkin, 1983).

Whether by causing a long-term change in the patterns of resource use and population dynamics of some species within the community, or by causing actual extinction, population interactions such as competition and predation, parasitism or herbivory may have potentially dramatic effects upon the composition and dynamics of ecological communities.

2.4 EMPIRICAL DEMONSTRATIONS OF COMPETITIVE EXCLUSION OR PREDATOR-PREY EXTINCTIONS

All the simple mathematical models we have introduced in this chapter for two-species interactions predict that in the extreme, competition and predation may result in extinction of one or both interacting populations. Potentially exclusive competitors might coexist in those situations where (i) resources were superabundant and thus actual competitive interaction absent, where (ii) conditions for both species were suboptimal and thus both were more strongly influenced by intraspecific competition than by interspecific interaction (e.g. Ayala, 1970) or more generally where (iii) the different species are competing simultaneously for a number of independent resources and differ in which particular resource most powerfully limits their population growth (Tilman, 1980, 1982) – and thus again, in the limit, are each more powerfully affected by intraspecific competition than interspecific interaction. But where interspecific interaction is strong, all models predict a real potential for extinction of one or other of the interacting species.

The potential for such extinction has primarily been demonstrated in practice in the laboratory in simple, species-poor systems in rather homogeneous environments. Thus the classic 'bottle experiments' of Gause (1934) on exclusion in competitive cultures of *Paramecium* (Figure

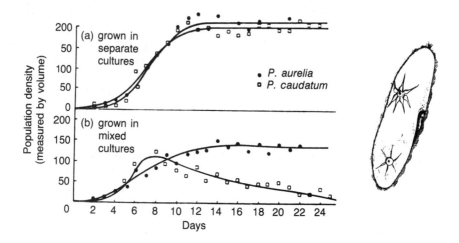

Figure 2.8 Growth of *Paramecium* populations (*P. aurelia* and *P. caudata*) in separate culture (a) and mixed culture (b). (After Gause, 1934.)

2.8 here), the subsequent studies of Park on flour weevils (Park, 1954; Park *et al.*, 1964) and of Ayala on *Drosophila* (Ayala, 1970) considered the dynamics of two species forced into strong competition in a homogeneous medium.

Sequential extinctions of first prey, immediately followed by its predator, in the simple one-predator one-prey system established by Huffaker between the orange mite *Eotetranychus sexmaculatus* and the predatory *Typhlodromus occidentalis* (Figure 2.9) were replaced by persistent oscillations in the numbers of both species when the complexity of the environment was increased by providing cocktail sticks at intervals among the oranges, from which *Eotetranychus* could launch itself in aerial dispersal to colonize 'new' oranges and establish for itself refuges ahead of the cursorial *Typhlodromus* (Huffaker, 1958). In the same way simple heterogeneities of the environment (complexities of either physical or biotic context) might prevent competition between two species from proceeding ineluctably to extinction.

Park's studies of *Tribolium* showed clearly that competitive superiority depended heavily on environmental conditions (with *T. castaneum* consistently outcompeting *T. confusum* under hot moist conditions but *T. confusum* having a clear competitive advantage under colder or more arid conditions (Park, 1954)). Other examples also show that the outcome of a competitive interaction may be determined in this way by environmental conditions (Ayala, 1970; Goldsmith, 1973); in Ayala's experiments

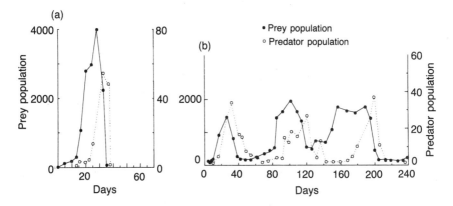

Figure 2.9 The effects of structural complexity on predator–prey interactions. Predator–prey interactions between the orange mite *Eotetranychus sexmaculatus* (closed circles) and its predator *Typhlodromus occidentalis* (open circles).
(a) In a structurally simple system a single oscillation is observed terminating in extinction of both predator and prey; (b) in a more complex system, sustained coexistence is facilitated by provision of prey refugia. (After Huffaker, 1958.)

with the fruit flies *Drosophila pseudoobscura* and *D. serrata*, *D. serrata* outcompeted *D. pseudoobscura* when culture temperatures exceeded 25°C; at temperatures of 22°C or below, *D. pseudoobscura* outcompeted *D. serrata*. In each case the two species might coexist in a more heterogeneous environment where different patches of the culture medium favoured one or the other.

2.5 COMPETITIVE COEXISTENCE: THE ROLE OF SPATIAL AND TEMPORAL HETEROGENEITY

Competitive advantage varies with environmental conditions; coexistence even of fierce competitors might thus be possible in a patchy environment where temporal or spatial variation in environmental conditions may result in corresponding alternation of competitive advantage of the competing species. That spatial heterogeneity does indeed facilitate the coexistence of competitors is illustrated by the work of Atkinson and Shorrocks (1981, 1984) with *Drosophila*, while the role of temporal variation in conditions is elegantly demonstrated in Spiller's analysis of seasonal reversals in competitive advantage between two orb-weaving spiders, *Metepeira grinnelli* and *Cyclosa turbinata* (Spiller, 1984). Coexistence of potentially exclusive species might equally be facilitated if competitive advantage varied at different stages of the life cycle (e.g. Haigh and Maynard Smith, 1972; Grace, 1985).

Hutchinson (1953) pointed out that the role of regular environmental fluctuations in permitting the coexistence of ecologically similar forms depends on the relationship between the periodicity of the environmental fluctuations and the generation time of the organisms concerned. If the generation time of the organisms is very short in relation to the period of environmental fluctuation, competition can occur in one direction over several generations within a single phase of the environmental cycle and may thus be able to run its course to the elimination of one or other species before conditions change. If the generation time is too long relative to the environmental change, any one individual must cope within its own lifetime with all environmental conditions; the better-adapted species may still outcompete the latter. However, where the relationship between environmental periodicity and generation time is intermediate, the environmental conditions may alternately favour one and then the other species for periods of a few generations so that both may coexist. In essence, coexistence under these conditions is permitted because competition is never allowed fully to run its course, and conditions are reversed before either one or the other species becomes extinct.

This same realization that apparent coexistence of fierce competitors within a community may result from fluctuations in environmental conditions of such a periodicity that competitive interactions are prevented from running to conclusion has led to the recognition of a number of other non-equilibrium models of competitive coexistence. The regular cycle in conditions envisaged by Hutchinson, favouring first one and then the other of a pair of competitors, is indeed rather a special case; even random, unpredictable variations in conditions, if of a frequency sufficient to disrupt the community's dynamics, may forestall competitive exclusion (see for example Connell, 1978; Miller, 1982; Chesson and Case, 1986). Huston (1979) has shown that the introduction of occasional episodes of heavy density-independent mortality into the classical Lotka–Volterra models of competition with which we began this chapter can delay almost indefinitely the final resolution of exclusion. Sale (1977, 1979) has introduced the element of random chance into competitive models by proposing a 'lottery model' of recruitment. This lottery model of competition was invoked by Sale to explain the coexistence of apparent competitors among reef fish assemblies (Sale, 1977, 1979; Sale and Williams 1982; Abrams, 1984; see also Gladfelter and Johnson, 1983), but is equally applicable in other situations where opportunities for reproductive recruitment or establishment of progeny within the community occur at random. Sale's model proposes that where the gaps needed for recruitment arise unpredictably through time and space and are filled on a 'first come first served' basis from a pool of potential propagules of all species, the first colonist to locate the vacancy (a vacant territory on a coral reef or a gap in a closed chalk grassland sward) will be able to hold the resource against later arrivals. Since it is a genuine lottery as to which species arrives first in any 'gap', no one species consistently wins, and potential competitors may coexist within the community. The effect of this is illustrated in Table 2.2, using some of Sale's original data for reef fish. Clearly such a model is only applicable where competition is for space or discrete resource patches, but it may have particular relevance in plant assemblages where, due to a generally sedentary habit, intensity of competition is greatest with immediate neighbours in space rather than more generally throughout the community.

Table 2.2 Lottery competition in coral reef fish

Resident lost	Reoccupied by		
	E. apicalis	Pl. lachrymatus	Po. wardi
E. apicalis	9	3	19
Pl. lachrymatus	12	5	9
Po. wardi	27	18	29

Three species of pomacentrid fish, *Eupomacentrus apicalis*, *Plectroglyphidodon lachrymatus* and *Pomacentrus wardi* colonize rubble patches of the Great Barrier Reef and establish exclusive feeding territories. The three species appear to coexist in competition with each other due to the fact that colonization of vacant territories appears a lottery. Data from Sale (1977, 1979) show for 131 'changeovers' the number of occasions when a territory originally held by one of the three species was successfully colonized by another individual of the same species or by an individual from one of the other species. Transition seems effectively random.

The importance of such non-equilibrium models of coexistence – and the role of stochastic variation in environmental conditions more generally in influencing the structure of communities – will be addressed in more detail in Chapter 7. In the meantime these various models and others are considered in an excellent review by Silvertown and Law (1987). Much of their review focuses on coexistence of competitors within plant assemblies since it is indeed amongst plants that the greater bulk of examples of apparent coexistence without clear niche differentiation may be observed; the factors involved and the processes reviewed are however equally applicable to any community.

2.6 SEARCHING FOR COMPETITION IN REAL COMMUNITIES

For all these reasons, while the potential for extinctions in either competitive or predatory relationships may be demonstrated in the relatively simple environmental contexts provided by laboratory culture, the incidence of actual exclusion in the more complex physical and biological environments of most natural communities is probably considerably lower than such potential might suggest. Indeed, while it is not especially difficult to substantiate the incidence of predation or parasitic interaction it is rather more difficult, as we shall see, to establish that serious competition is ever experienced in natural communities. The difficulties are two-fold.

In the first instance, a great deal of evidence is necessarily circumstantial. Much of the evidence adduced in support of some major role of competition in influencing community structure is based on (i) observations of regularly repeated patterns of co-occurrence of particular species sets of low average overlap in resource use (and conversely, the apparent absence of otherwise equally plausible species combinations, which would result in higher overlaps in resource use (Diamond, 1975 *et seq.*, Chapter 6), or on (ii) observed differences in resource use patterns of a single species in different contexts – in one case exploiting its preferred resources in relative isolation, in the other case forced to share those resources with other species (**niche shifts**; Chapter 4). In both cases we are forced to infer the role of competition in retrospect; we cannot prove that the patterns we perceive are the result of competition; we are merely presented with a *fait accompli*. Further it is suggested by many that the alternative approach, a search for clear examples of competition within present day assemblages, is equally unlikely to resolve one way or the other the debate about the potential role of competition in structuring communities; for almost by definition, maximum efficiency should involve avoidance of interspecific competition to escape the inevitable loss of fitness consequential upon such interaction. Thus, while organisms may respond to the potential for competition, actual competition may rarely be encountered in natural communities; and in the response to a potential threat it is far harder to attribute causality. Such arguments have led some to postulate that interspecific competition will rarely if ever actually be expressed in natural communities; or if it is, it will be so ephemeral in its expression that it will rarely be detected. Yet despite the fact that conscious search for such examples might seem like looking for a needle in a haystack, occasional chance observations do arise – and manipulative experimentation may reveal underlying dynamics of competition even in systems apparently in equilibrium.

Interpretation of such observations and experiments must however be undertaken with extreme caution. Reynoldson and Bellamy (1970) have published a set of essential criteria for imputing the existence of competition from simple field observation, while in an excellent critique of experimental manipulations Bender, Case and Gilpin (1984) distinguish clearly between two distinct types of manipulation and establish criteria for determining both the appropriate protocol to adopt in any set of conditions, and the interpretation of the results of both PRESS and PULSE manipulations. (PULSE perturbations are defined as a momentary intervention or alteration of species numbers by the experimenter, after which the system perturbed is allowed to relax back towards its previous equilibrium; a PRESS manipulation involves study of the consequential dynamics of the system when some artificial alteration of species number within the community is maintained over a longer period; Bender, Case and Gilpin, 1984.)

One of the most fortuitous – and compelling – examples of competition and its consequences in unperturbed systems is the documentation by Reynoldson and Bellamy (1970) of the sequence of events following the chance colonization by the flatworm *Polycelis tenuis* of a freshwater lake previously inhabited only by the congeneric *P. nigra*; Reynoldson and Bellamy showed that in response to colonization of the lake by *P. tenuis*, populations of *P. nigra* were reduced to become only some 10% of the total triclad population of the lake. Since the total *Polycelis* population remained more or less unchanged, there must have been actual replacement of *P. nigra* by *P. tenuis* (Reynoldson and Bellamy, 1970).

Such chance observations are, however, far from commonplace and the search for evidence of competition has more generally relied upon interpretation of manipulative experiments in which one species is removed from or added to an established community and the responses of the other species within the system are monitored over time. Both Connell and Schoener published in 1983 analyses of all the published accounts that they could find in the literature testing for interspecific competition in field experiments. Of the 164 studies reviewed in Schoener's analysis, roughly equal numbers had focused on terrestrial plants, terrestrial animals and marine organisms; fewer studies had been undertaken in freshwater systems, and among the studies undertaken within terrestrial systems, surprisingly few had looked at phytophagous insects. Between them the various studies had looked for evidence of competition amongst 390 pairs or groups of species (many of the studies dealing with several groups of species); 76% of species showed the effects of competition at least sometimes, with 57% showing clear evidence for competitive interaction under all conditions.

Seventy-two studies were reviewed by Connell (1983) dealing with 215 species in a total of 527 different experiments. Interspecific competition was demonstrated in more than half the species, and in approximately 40% of the experiments (Connell, 1983). Although both reviews certainly confound those who would claim that competition is rarely if ever important in influencing the structure and dynamics of real-world communities, it is only fair to note that both indubitably also overestimate the incidence of competition in the real world. All the studies quoted involved deliberate manipulations to 'provoke' competition. Further, many pairs of species studied were probably selected for study because competition was suspected: in this sense they may not offer a truly representative sample. In addition, the studies published were equally probably accepted for publication because they did show something interesting: like newspapers or news broadcasts, journals less commonly publish non-stories. Finally, there is another kind of bias in the results: that the relative contributions to the data of different types of organisms and different taxonomic groups were far from equal. As already noted, (and noted by Schoener himself (1983)) phytophagous insects in particular were surprisingly under-represented in his data – despite the fact that they account for roughly a quarter of all living species! And, as we shall see (page 111), it seems probable that interspecific competition is in fact far less prevalent among such assemblages; in a review of 41 published studies, Lawton and Strong (1981) and Strong, Lawton and Southwood (1984) found evidence for competition in only 17 assemblages, and in those the evidence for actual competition was restricted to a very small proportion of the possible pair-wise combinations of species.

Of the many published examples of field manipulations, my own particular favourites include the studies of Hairston (1980) and Menge (1972, 1976). Hairston has demonstrated for two sympatric species of terrestrial salamanders, *Ptethodon*, in the Great Smoky Mountains and the Balsam Mountains of North Carolina, that experimental removal of *Ptethodon jordani* from marked plots resulted in a statistically significant increase in abundance of the sympatric *P. glutinosus*. Removal of *P. glutinosus* (the less abundant species) did not result in an equivalent increase in numbers of *P. jordani*, but did increase markedly the overall proportion of young individuals within the population. In another field experiment, on competition between two sympatric species of starfish, *Pisaster ochraceus* and *Lepasterias hexactis*, Menge removed all the individuals of *Pisaster* from one small island reef and added them to another similar reef elsewhere, effectively doubling their density in the second site; a third reef nearby was left undisturbed to act as a control. Subsequent monitoring of the starfish showed that while the size of control animals remained unchanged, the average weight of individual *Lepasterias* increased significantly with removal of *Pisaster* and decreased with its addition. Once again this is highly suggestive of some competitive interaction, a conclusion supported by the additional observation that the standing crop of *Lepasterias* (biomass per m^2) on undisturbed reefs was inversely correlated with that of *Pisaster* (Menge, 1972).

Pisaster ochraceus is involved in yet another example which may be advanced here as evidence for competition. This starfish is the dominant predator in an intertidal system of some 16 rocky shore invertebrates; many of these species overlap broadly in feeding ecology but seem to be able to coexist without evident competition. However, when *Pisaster* was artificially removed from the system, in an elegant series of experiments by Paine (1966), the species number within the prey community fell dramatically: of the 11 species fed upon directly by *Pisaster*, seven

Diffuse competition, mutualism, indirect effects

disappeared from the community altogether, to leave a new community of four species from the previous array fed upon by the starfish and four non-prey species. This can surely be attributable only to the effects of interspecific competition among the species previously preyed upon by *Pisaster*, competition which was suppressed in the presence of the dominant predator. With the removal of *Pisaster*, prey species could increase in numbers to levels higher than those observed in the presence of predation; resources became limited and competitive dominance from mussels (*Mytilis edulis*) resulted in the exclusion of many species of the original prey assemblage.

This last example is by now such a classic that it risks being over-quoted. Yet it is important, not just for the evidence presented for the potential effect of competition within real world communities, but because it carries implicit within it illustration of another crucially important point: that interactions between any given pair of species within a community may be markedly affected by other, external, interactions of the community involving one or other, or both of those organisms.

2.7 INDIRECT EFFECTS: DIFFUSE COMPETITION, COMPETITIVE MUTUALISM AND OTHER MORE COMPLEX INTERACTIONS WITHIN MULTISPECIES SYSTEMS

So far in this chapter we have primarily considered direct interactions of one species with another; in addition we have been dealing almost exclusively with the implications of such interactions for clearly defined pairs of protagonists. Yet under natural conditions such interactions are necessarily set within the context of others occurring all around them – and it is clear that they are more effectively viewed within such context: the apparent consequences of any direct interaction may be modified or masked by the effects of other interactions involving the same organisms.

We have noted (page 31) that coexistence of strong competitors or single predator–single prey systems may be facilitated by increasing the physical complexity of the environmental 'arena' in which that interaction is staged. In the same way the biotic complexity of the context may also affect the outcome. Thus, as we have already noted, competitive coexistence may be facilitated by the presence within the community of a generalist predator, or other consumer (effectively restricting the population sizes of all potential competitors below the level at which resources become limiting), or where predation impinges selectively on the apparently superior species of some competitive interaction (Paine, 1966, 1969). Similar results to those reported by Paine for intertidal communities of rocky shores in California are documented amongst algal communities in southern Chile: Jara and Moreno (1984) note that herbivore removal from the mid-littoral community resulted in high percentage cover of available rocky surfaces (>80%) by the red alga *Iridaea boryana*; herbivore addition or a natural increase in herbivore densities resulted in a dramatic decline in percentage cover of this alga and the substrate became dominated by barnacles and other, crustose algae (Jara and Moreno, 1984).

Indeed in terrestrial as much as in intertidal systems, herbivory is widely implicated in such **competitive release**; grazing by the sheep flocks of the medieval graziers – and subsequently by rabbits – is widely held to be responsible for the tremendous floristic diversity of English chalk grassland (Tansley, 1922; Tansley and Adamson, 1925; Hope Simpson, 1940). By reducing the vigour of the taller-growing grasses and maintaining a short sward of close-cropped turf, grazers prevented the grasses from growing to such a height that they might exclude other lower-growing species from the community through shading (competition for light) and thus helped maintain an unusually high diversity of shade sensitive species within the community.

Additional competitors for one or other species of some interaction may have a similar effect – in suppressing populations of a strong interactor, such 'external' competitors may restrict the effects of competition on a weaker third party: a phenomenon referred to as **competitive mutualism**. Equally, the presence of other competitors within the community, even if only interacting weakly with each of the main protagonists considered, may enhance the effects of the direct competition; such **diffuse competition** may have a major contributing effect to the observed response of the species considered. Finally, extended interactions not taken into account may also lead to false interpretation of the interactions which are considered: Holt (1977) argued that two species might mistakenly be construed as competitors when in fact their negative interactions were the consequence of an (unconsidered) shared predator. Holt termed this relationship **apparent competition** because in the absence of the predator, the two species did not compete.

The suggestion that the effects of any primary interaction might be moderated to some extent by the effects of other interactions spreading out beyond the limited arena on which we have focused must lead us to explore whether we may reassemble such binary interactions into a more global model of the community as a multispecies complex. We have already noted that the models introduced on pp 15–21 may be extended to embrace interactions between three-, four-, five-, n-species sets.

Such extension from the simple analyses of binary interaction leads to perhaps more representative analysis of the interactive dynamics of complex, multi-species assemblies: studies of the dynamics of the individual component populations within more complex webs, the types and

Diffuse competition, mutualism, indirect effects

strength of interaction, the topology of linkage within such a web, and the number of interactions in which any one organism may be engaged (Chapter 3). Alternatively, we may extend our simple analyses of the effects of competition or predation on patterns of resource use of single pairs of interactors, to consider the assortment of multi-species sets within resource space (Chapters 4 and 5).

3
Food webs and connectance

One of the most productive areas of recent research into community structure and the determinants of such structure has been the analysis of food web patterns. While restricted in its coverage to the analysis primarily of heterotrophic feeding relationships within communities, analysis of community structure at the level of the food web does at least permit recognition of at least some of the complexity of a multi-species system of interaction.

We introduced the approach briefly at the end of Chapter 1. In essence food web studies depend upon the development of computer models of interacting populations of random assembly, whose dynamics are modelled generally by Lotka–Volterra mathematics (Lotka, 1925; Volterra, 1926; and see May, 1973a). The dynamics, persistence and stability of such populations are examined over a number of generations and the common characteristics of that subset of models which do show some measure of stability identified by comparison to the parameters defining unstable model webs. Finally, the characteristics of these stable model webs are compared with the properties of real life systems. An overview of the approach is presented by Paine (1980) and a summary of the main conclusions of such analyses is offered by, for example, Pimm (1980a, 1982) or by Pimm, Lawton and Cohen (1991).

It is emphatically not the intention of this book to offer a detailed analysis of the mathematical formulation of such food web models, nor their interrogation. The reader interested in delving deeper into the underlying theory would do well to refer to more specialist works such as Paine (1980, 1988), or Pimm (1982, 1991). It is my intention to focus rather on the major conclusions arising from such analyses and explore their implications in our search for patterns or rules in community organization.

Food web analyses of this type have highlighted a number of common properties of stable webs. Summarizing the main conclusions to have arisen from studies of food web design by the time of his review in 1980, Pimm noted:

> On the assumption that systems of interacting species, when perturbed from equilibrium should return to that equilibrium quickly, one can predict [a number of major] properties of food webs: (1) food chains should be short; (2) species feeding on more than one trophic level (omnivores) should be rare; (3) those species that do feed on more than one trophic level should do so by feeding on species in immediately adjacent trophic levels (Pimm, 1980a)

To such a list we might now add a recognition of restrictions on the topological structure of the web – the nature and direction of certain minimal linkages required to provide a mathematically rigid structure (Sugihara, 1984). Stable food webs are characteristically 'upper triangular' (Figure 3.6); that is to say, if species from the web are arranged into a simple matrix of consumers and consumed (with all species present recorded in both rows and columns as at least potential predators and prey), actual feeding relationships all lie to one side of the central diagonal: all feeding relationships are consistently transitive (Cohen, Newman and Briand, 1985; Cohen, Briand and Newman, 1986). Finally, studies of the dynamics of both model and real world communities suggest very restrictive constraints on the number of links 'permitted' in a web of given species number.

3.1 IMPORTANT ASSUMPTIONS

A number of the conclusions of food web studies are dependent upon assumptions implicit in the community models employed and it is important at this stage to consider exactly what those assumptions are – and what may be their implications. Thus all food web models are based upon the Lotka–Volterra differential equations introduced in Chapter 2, extended to a more general multi-species form (page 17). Interaction coefficients between each pairwise combination of species within the array are summarized in a 'community matrix'; for each pair of species the type and strength of interaction are represented by the magnitude of the α-values recorded within the matrix for that particular combination of species and the signs associated with each coefficient (++, +–, ––, +0, 00, etc. (Table 2.1)).

While Lotka–Volterra models of this kind impartially reflect competitive mutualistic or neutral interaction, +– interactions are usually presumed to be 'top-down': acting in such a way that the abundance of predators influences strongly the dynamics of their prey but are themselves relatively independent of changes in the abundance of those prey. (It is assumed, if you like, that if one particular prey becomes scarce, the predator can maintain its own population levels by switching to alternative prey species.) By contrast, certain types of population interaction might be 'donor-controlled': changes in population size of a prey may be relatively independent of the population size of its predator, while the predator's dynamics depend more critically upon the availability of its prey. It is generally believed that donor-controlled relationships are relatively rare in predator–prey, herbivore–plant interactions in the real world – that in effect few plant or animal populations are unaffected by the abundance of their consumers; most food web models are therefore based upon a presumption of top-down interactions. But we should note that decomposer systems are not uncommonly donor-controlled (Pimm, 1982; Hildrew, Townsend and Hasham, 1985) and the relative contribution of top-down and bottom-up forces in community ecology more generally have recently been re-examined by Hunter and Price (1992), Matson and Hunter (1992), Menge (1992), and Strong (1992). As noted by Pimm (1984) and many others, food webs assembled with a presumption of donor-controlled dynamics may have very different properties.

3.2 TROPHODYNAMIC IMPLICATIONS OF FOOD WEB STRUCTURE

Some of the earliest explorations of the dynamics of model food webs had immediate relevance to the trophodynamic model of community structure developed by Elton, Odum and others. It quickly became apparent that certain trophic relationships within a community tend to be destabilizing: that is, for example, communities become progressively less likely to persist as the degree of omnivory within them increases (species which feed upon more than one other trophic level; Pimm and Lawton, 1978). From such analyses, Pimm and Lawton predicted that food webs with high numbers of omnivores will be rare in the real world – a prediction supported by data on real communities assembled by Cohen and others (Cohen, 1978; Briand, 1983; Cohen, Briand and Newman, 1990). As a rather pleasing exception, omnivory, while rarely encountered in consumer webs, is far more commonly recorded in decomposer systems. And, in the spirit of exceptions proving the rule, it is such webs, as we have noted, which are more likely to show donor-controlled dynamics.

Further analyses of model webs confirm an upper limit to the number of trophic levels which may be supported in any given system (page 7) but offers an alternative derivation for such a conclusion. Pimm and Lawton (1977) argued that limits to the number of trophic levels within a system (or links in any one food chain) would result necessarily from consideration of the dynamical stability of such webs. While their evidence is empirical, based on the repeated analysis of the stability properties of model webs, we may try to visualize the argument at a more intuitive level (although we should note that such intuitive presentation may not represent the true logic with absolute accuracy).

In essence, studies of the dynamics of any single-species population show that such populations very rarely display stable point equilibria, but at best express some form of damped oscillation around a stable point. Deviations from the equilibrium point are restricted and quickly corrected – but larger deviations may result from time to time in response to stochastic variation (changes in carrying capacity, periods of high density-independent mortality, etc.). Thus the dynamical behaviour of even a single population in isolation is subject to substantial variation.

Such stochastic variation is amplified in interaction with other species, within two- or three-species model systems. As we have seen, both predation and competition as biological processes tend to be destabilizing and have a disrupting effect on population dynamics of the participants (although predation can stabilize a competitive interaction that would otherwise be unstable and the effects of predation in general do depend very heavily on the context). Random oscillations in population size of some prey species about some equilibrium level will be amplified within the population fluctuations of a predator. The predator may be feeding on a number of alternative prey species and by random chance the population sizes of all its prey may simultaneously be freakishly low, or may simultaneously peak; such synchronous aberration would be additive in its effects and would be reflected in more extreme fluctuations around its mean value for the population of predators. Stochastic fluctuations in population size would be accumulated up the food chain, with less and less predictable behaviour of populations higher up the trophic scale. Eventually the unpredictable dynamics of a top predator would be such that it would be impossible to maintain a population at a higher level still with any semblance of stability – that any minor perturbation to the system would be sufficient to displace it.

Elton's initial observation that in natural communities food chains were generally limited to three or four trophic levels had historically always been explained by arguments based upon the inevitable attenuation of energy (Hutchinson, 1959; Slobodkin, 1961). Pimm and Lawton's observations offer a totally different explanation: that food chains will be restricted in length purely as a consequence of the dynamical behaviour of populations of higher order predators. Which theory may be regarded as more 'correct'?

Pimm and Lawton suggest that arguments based on attenuation of energy are fatally flawed; on such a basis one should expect that communities with higher levels of energy input or GPP (e.g. Figure 1.2) should be able to support a higher number of trophic levels overall than those systems (like arctic or alpine ones) with low energy availability. Even though the energy loss at each step is proportional to the total energy available in that trophic level, nonetheless communities with initial energy inputs higher, perhaps by orders of magnitude, should in general be expected to show a capacity for supporting a higher number of trophic levels than more 'impoverished' ones. I confess that at a personal level I have never found this rather nebulous, 'armchair' type of reasoning particularly convincing. However, more recently these arguments have been better validated in an elegant experimental test by Pimm and Kitching (1987).

Pimm and Kitching established experimental 'communities' equivalent to those found in natural tree-holes by setting out water-filled pots on the forest floor in rainforest communities of southern Queensland. These artificial phytotelmata were exposed to natural colonization, but levels of available energy were manipulated to establish high energy and low energy systems, by addition of different quantities of leaf litter. Food chain lengths were not directly related to energy availability: one top predator (tadpoles of *Lechriodus fletcheri*) actively avoided the more productive treatments. The other main predator (larvae of the chironomid midge *Anatopynia pennipes*) eventually colonized the majority of pots, whatever the initial energy levels; most significantly however, it was slow to colonize the experimental pots, establishing itself only after its preferred prey species had reached approximately constant population size. Pimm and Kitching suggest that these data indicate that frequent disturbances would remove the larvae from the system. Comparative studies of the variations in food chain length in natural phytotelmata, in relation to available energy in the form of detritus inputs, also suggest that the susceptibility of top level predators to disturbance may indeed be the major factor determining food chain length (Pimm and Kitching, 1987).

3.3 INTERNAL STRUCTURING WITHIN FOOD WEBS: PROBLEMS OF CONNECTANCE

Different links within the web clearly have a different relative importance in maintaining community structure and function, depending on the type of interaction and the interaction 'strength'; thus it is not merely the shape or topology of the web which will affect its dynamics and stability, but the distribution of stronger or weaker linkages, the direction and the absolute number of those links.

Patterns of linkage within webs may be characterized in terms of their degree of connectivity, or connectance, and average interaction strength. Connectance (C) may be defined simply as the number of links represented within a real or model web, as a proportion of the number topologically possible (or $L/S(S-1)$, where S is the number of species within the web and each could theoretically be connected to every species other than itself). The relative 'strength' of any interaction may be determined in relation to the degree of displacement of the population trajectory of one of a pair of interactors resulting from a given manipulation of the population size of the other member of the pair; average interaction strength within a web is variously denoted b or i in different published analyses.

A number of early studies (reviewed May, 1981, 1986a) noted that the stability of model webs (as the period to extinction) remained relatively constant when the term, $b(SC)^{1/2}$ remained below 1. As values approached or exceeded unity, such systems suddenly became unstable, almost as if approaching a catastrophe plane. Although Cohen, Briand and Newman (1990) have since suggested this particular formulation may be algebraically incorrect (at least in the generality claimed), it is certainly a general phenomenon that large and complex systems – whether physical or biological – may be expected to show increased stability with increased complexity up to a critical level of connectance, and then, as connectance continues to increase, to go suddenly unstable (Gardner and Ashby, 1970).

In systems of given interaction strength, b, such a relationship essentially reduces to the observation that in stable communities the product SC (Species number × Connectance) should be effectively constant; $SC = k$. If this relationship is a necessary requirement for stability – and if there is some form of catastrophe plane when $b(SC)^{1/2}$ approaches unity, fulfilment of this constraint implies a dynamic trade-off between species richness of a community and connectance, such that connectance must decline as species number increases.

In a critical evaluation of this and other community patterns, May (1986a) confirms that an inverse relationship between S and C does appear to be characteristic of stable models and that the relationship also appears to hold in real-life assemblies. Thus in analyses of plant-aphid-parasitoid communities in central Europe, in which species richness ranged between 3 and 60, Rejmanek and Stary (1979) found that connectance did indeed vary inversely with species richness, with SC approximately constant at around 3 (in practice values ranged from 2 to 6) (Figure 3.1). Yodzis (1980) has analysed some 24 food webs catalogued by Cohen (1978) – webs exhibiting a far greater range of kinds of interaction than the more homogeneous plant-aphid-parasitoid systems of Rejmanek and Stary; in his analyses too, SC remains more or less constant (with $SC \simeq 4$, as S ranges from 8 to 64). Finally, McNaughton has collected data on plant species in 17 different grassland assemblages in East Africa (McNaughton, 1978); for such assemblages, in which, by contrast, all interactions were by definition competitive, the product SC remained remarkably constant at between 4 and 5.

May notes, however, following Pimm (1980b), that the relationship

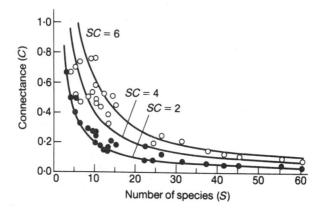

Figure 3.1 The relationship between Connectance (C) and Species richness (S). The empirical relationship between C and S determined by Rejmanek and Stary (1979) in a series of plant-aphid-parasitoid communities is represented. Closed circles show minimum estimates of connectance, open circles maximum estimates in each case. Expected curves for constant product SC = 2, 4 and 6 are superimposed

between S and C is essentially trivial and need not necessarily reflect any dynamic interaction between connectance and species richness, or any necessary trade-off (May, 1986a). $S \times C$ will always remain constant in any system as long as any organism within that system has a limit to the number of links it would normally make with other species around it and provided that each additional species added to that system only brings with it about the same number of new linkages (i.e. does not result in an increase in the average number of links per species) (Box 3).

May also quotes Briand and Cohen (1984) and Cohen and Newman (1985) in analysis of 62 real webs to show that linkage density is approximately constant (at ~ 3.7). Briand (1983) and Cohen, Newman and Briand (1985) also suggest, however, that linkage density is systematically higher for food webs in constant environments than in fluctuating environments: thus for given S, connectance is higher in constant environments than it is in more variable environments. Such an observation accords well with the notion (Chapter 9) that more complex communities tend to be able to develop under conditions of greater environmental constancy (May 1972, 1973a *et seq.*). Further, Margalef and Gutierrez (1983) observe that connectance is positively correlated with rates of turnover within a community (measured as Primary Production/Community Biomass), suggesting that more productive systems may have higher connectance.

All such rationalization, and reduction of the observed relationship between species richness and connectance within a system to a requirement of constant linkage density, merely replaces the original question

Internal structuring within food webs

with a new and fundamental puzzle of why should the number of interactions involving each species be so nearly constant. And why is the average number of interactions for each species so low (~ 3.7 in most systems)?

BOX 3

The product SC (Species richness × Connectance) appears to be approximately constant in stable model webs – an observation apparently born out in empirical studies. However Pimm (1980b) and May (1986a) both suggest that such a relationship between S and C is essentially trivial and need not necessarily reflect any dynamic interaction between connectance and species richness; S × C will always remain constant in any system as long as the average number of links per species, or **linkage density** is constant over all systems. This is relatively simply established:

Let Connectance (C) be defined as on page 44 as the total number of observed links within a web (L) as a proportion of the total number of links theoretically possible (S(S–1))

L is the total number of links in the web between pairs of species; we may thus also calculate the average number of interactions for each species (or linkage density, d) as: L/S

Thus:

$$C = \frac{L}{S(S-1)} = \frac{S.d}{S(S-1)}$$

If S is reasonably large, (S–1) tends to S; and C tends to d/S
With $C \simeq d/S$, product $SC = k$ reduces to $d = k$

(An alternative proof may be offered that, for product SC to remain constant, C must decrease at the same rate as S increases – in a hyperbolic function (as Figure 3.1). Using the same notation as above:

$$C = \frac{L}{S(S-1)} \text{ tends to } \frac{L}{S^2} \text{ when S is large}$$

For C to decrease as S increases L must scale with S. But L = S.d and therefore L will indeed scale with S as long as d, the linkage density, remains constant.)

In such analyses the relationship $SC = k$ in itself becomes mathematically trivial, as Pimm and May suggest; its implications are merely that:
1. whatever the system, more or less, the average number of links per species is constant (or at least as constant as k: perhaps 3–5).
2. within any system, or between systems, an increase in S can only be accommodated if each new species adds to the system no more than the 'average number of links per species': i.e. any new species entering the system must not alter linkage density, d.

More recent analyses of connectance within food webs have challenged two of the initial premises:

1. that linkage density (or product SC) is constant in real systems (Paine, 1988; see also Winemiller, 1989);
2. that d (the number of links per species) will remain constant as S increases (Warren, 1990; Cohen et al., 1990).

In an excellent critique of food web theory, Paine (1988) notes that both the apparent constancy of product SC and its low recorded value (~ 3.7) may both be artefacts of levels of perception/resolution in the analysis of real food webs. Paine argues that the various published food webs used to test the predictions of theoretical analyses (compilations published by Cohen (1978), Briand and Cohen (1984), Cohen and Newman (1985), and Cohen et al. (1990)) vary tremendously in quality and resolution. Few, if any, were collected primarily for the purpose to which they are now put in validating apparent structural constraints on web design. Paine points out that the published studies often vary in level of taxonomic resolution – not just in separate analyses of web structure, but even within different regions of a single web – and that calculated values of connectance may vary depending on the level of resolution or aggregation of species.

(As illustration, Paine calculates a variety of different connectance values from Menge and Sutherland's (1976) description of the food web of invertebrates on rocky shores of New England. Depending on whether three species of winkles (*Littorina*) are considered separately or as a single entity, calculated values of connectance, C, change from 0.31 to 0.36; if a crab (*Carcinus*), fish (*Tautoglobarus*) and a complex of vertebrate predators are added to the web, as suggested by Edwards, Conover and Sutter (1982), C increases to 0.44; when the vertebrates are disaggregated to birds-mammals-fish, and the three *Littorina* species are distinguished separately, $C = 0.46$.)

Aggregation is particularly common at the bottom of such webs, where species are smaller, less observable and commonly less well-identified: Paine cites the marine community described in Briand's (1983) matrix 29, where top species are individually treated ('right whale', 'bearded seal', etc.) while lower orders are extensively aggregated ('clupeid fishes', 'benthonic invertebrates'); many authors offer generic groupings such as algae, detritus, plankton. Such a procedure must obscure a number of independent linkages and minimize the number of prey categories recognized.

Paine further notes that any recording of a web of interaction is necessarily incomplete – and restricted by simple human limitations. Not all species/interactions are readily observable; nor is that proportion likely to be a constant. When S is small, a higher proportion of the links may be discovered; when S is large, the sheer complexity of the web suggests that a larger proportion of possible links will be overlooked. Yet other species may be observable but residence time within the study area may be limited due to their mobility, so that the impact of, for example, birds, mammals or other larger vertebrates on a web's dynamics is probably underestimated. It is even possible that in the recording of interactions within new webs currently being documented, the observer notes the immediately obvious links and a proportion of the more subtle interactions, but then subconsciously stops his search among the obscure linkages when connectance approaches the 'correct' theoretical value. In this way unconsciously biased sampling might perpetuate the myth – and the apparent constancies of pattern.

Sugihara *et al.* (1989) have addressed this problem in some measure by examining the effects on several food web properties of progressively 'collapsing' the web into 'trophic species' (as Briand and Cohen, 1984) sharing the same predators and prey. This procedure groups web elements on a purely functional basis and without reference to taxonomic affinities. Hall and Raffaelli (1991), in their classic analysis of trophic relationships among the various organisms of the Ythan estuary near Aberdeen, have adopted the alternative device of progressively collapsing the Ythan web by grouping taxonomically related species (Table 3.1). Sugihara *et al.* (1989) found that most web properties were remarkably robust when webs were progressively collapsed, concluding that the variable quality and resolution of published webs would not significantly confound analysis of web statistics. Hall and Raffaelli suggest that while most web properties were indeed insensitive to progressive taxonomic clumping, determined food-chain lengths were not – with average apparent food chain length becoming progressively shorter as the level of resolution of 'species' within the web declined.

The question of whether or not we should **expect** product SC or the number of links per species to remain constant, even in well-documented webs, is raised by Warren (1990) who claims, by contrast, that we should expect the number of links per species to increase, one to one, as S increases.

Warren argues that if any predator has a limited range of conditions or resources over which it can operate, then the more 'prey species' added to the system, the more will fall within each ecomorphological range; hence the number of links per species increases directly as S increases. Such an argument, however, overlooks the potential for competition – and its consequences. Increasing overall the number of species within the web, while it increases the number of potential prey, increases also the number of potential competitors within its own trophic level and may reduce its own 'sphere of operation'. The proportion of resource space occupied by any predator may itself therefore decline as the

Table 3.1 The web structure of the Ythan estuary can be considered at different levels of taxonomic resolution

Level 1	Level 2	Level 3
Otter		
Cormorant		
Heron		
Duck		
Swan		
Wader		
Gull		
Tern		
Passerine	Otter	
Salmonid	Cormorant	
Clupeid	Heron	
Gadoid	Duck	
Flatfish	Swan	
Anguillidae	Wader	
Pipefish	Gull	
Sand eel	Tern	Piscivore
Butterfish	Passerine	Avian herbivore
Zoarces	Fish	Avian benthivore
Goby	Malacostracan	Fish
Bullhead	Mysid	Epibenthic crustacean
Stickleback	Copepod	Detritivore
Crab	Barnacle	Planktonic herbivore
Prawn	Polychaete	Sedent. planktivore
Amphipod	Oligochaete	Polychaete predator
Isopod	Mollusc	Benthic herbivore
Mysid	Meiofauna	Molluscan predator
Calanoid	Insect	POM
Barnacle	POM	Algae
Lugworm	Green algae	
Rag worms	Brown algae	
Paddle worm		
Spionid		
Capitellid		
Sabellid		
Aricia		
Hesionidae		
Parathemisto		
Nototropis		
Oligochaete		
Gastropod		
Bivalve		
Littorina		
Opisthobranch		
Harpactacoid		
Ostracod		
Nematodes		
Foraminifera		
Insect		
POM		
Green algae		
Brown algae		

POM denotes Particulate Organic Matter. (From Hall and Raffaelli, 1991.)

number of species in the web increases, through competition within its own trophic level.

> **BOX 4**
>
> The question of whether or not we should **expect** product SC or the number of links per species to remain constant, is raised by Warren (1990). Warren claims that we should in fact expect the number of links per species to increase, one to one, as S increases, with
>
> $$L = P(S(S-1))$$
>
> Warren argues that if any predator has a limited range of conditions or resources over which it can operate, then the more 'prey species' added to the system, the more will fall within each ecomorphological range; hence the number of links per species increases directly as S increases. Such an argument, however, overlooks the potential for competition and its consequences. The proportion of resource space occupied by any predator may itself **decline** as the number of species in the web increases, through competition within its own trophic level. If therefore
>
> P is proportional to $1/S$, for increasing S, $PS \sim$ constant (k);
>
> the derived relationship $L = S.d$ (from Box 3) = $(P.S(S-1))$, reduces simply to a reaffirmation of constant linkage density: $d = k$.

Despite such arguments, however, one suspects that there is an element of truth in Warren's assertion that the number of links should increase with species number. After all, the idea that linkage density might be constant is to an extent assumed by derivation from the original observation that product SC might remain approximately constant – an assumption which while based on reasonably robust theory and apparently supported by empirical evidence is now more widely challenged (Paine, 1988). Intuitively, too, one might imagine that the number of links engaged in by a species should increase slightly as S increases, purely because the opportunity for such linkage increases. The number of potential links must be restricted at low S, purely and simply limited by the availability of other species to link to; as S increases beyond the level where links are constrained by lack of availability of interactors, the number of links per species must, at least initially, rise with increasing S. Evidence for such an increase in linkage density with increased species richness within a web is more formally presented by Cohen et al. (1990).

And the implications of all this for product SC? SC should still remain approximately constant over a range of S. But, as species richness rises, while connectance will decline in concert, it will not decrease at quite the same rate. In consequence, as S increases, C will be slightly higher for any given S than would be predicted by simple hyperbolic decline, and product SC will increase slightly with S. Perhaps this may explain the observed variation in SC found in natural communities and the higher connectance observed in more predictable environments (Briand and Cohen, 1983). Greater environmental stability would permit the development of higher species richness (May, 1973a, 1975a) which would in turn permit development of slightly higher overall complexity, as SC.

3.4 COMPARTMENTS IN FOOD WEBS

Observation of an apparent trade-off between species richness and connectance led to speculation that this might reflect some form of internal structuring within the web itself, such that larger food webs might be organized as sets of small subunits of intense interaction linked only 'loosely' to other subcompartments. Such a suggestion would provide an obvious mechanism for keeping SC approximately constant within larger assemblages, and also ties in with the theoretical conclusion of May and others (May, 1972a; McMurtrie, 1975, Goh, 1979) that, for specified S and C dynamic stability of model food webs might be enhanced by some such substructuring within the web. Pimm and Lawton (1980) have re-examined both theoretical and empirical analyses; using different models, which they claim incorporate biologically more meaningful assumptions, they conclude that there is no necessary presumption that food webs are more stable if divided into blocks in this way. Further, in a review of published food webs, Pimm and Lawton conclude that there is no evidence for the general existence of such compartmentation in natural communities. Where subcompartments are apparent within such natural webs they seem to be primarily a function of habitat boundaries: reflecting the distribution of the community between different habitats or microhabitats; they claim there is no evidence for subcompartmentation within habitats. Although Raffaelli and Hall (1992) have subsequently claimed that there is compelling evidence for compartmentation in several of the benthic webs analysed by Pimm and Lawton (1980), Pimm and Lawton themselves concluded that there are at present neither adequate theoretical nor convincing empirical grounds for believing that food webs are divided into compartments. There may, however, be other structural constraints that stable food webs must obey.

3.5 FOOD WEB TOPOLOGY

While the separate interactions between pairs of species within a web clearly influence the dynamics of the whole, the structure of such links within the web as a whole (presence or absence of specific links within the matrix, position and direction of such linkage), may also be shown

Food web topology

to have an effect on web stability. Although studies of the topological structure of food webs are still rather rudimentary, the work of Sugihara (1984) and Auerbach (1984) has hinted at certain requirements of structure for stability.

Such analysis considers the links present (as linkages between resource consumers and those consumed) as a subset of all linkages possible within a web. Sugihara (1984) creates a first order derivative of such a web, creating (after Cohen, 1978) a **consumer overlap graph** where whenever two consumers overlap in their use of resources an edge is drawn between them. The vertices in the resulting model, or digraph, represent the different consumers (e.g. Figure 3.2) while the lines connecting them signify overlap in consumer resource use. In analysis of such models,

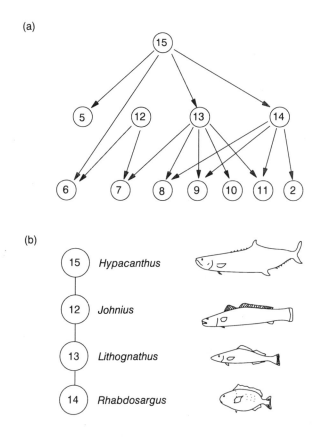

Figure 3.2 Consumer-overlap graph for the Knysa Estuary. In (a) a simple food web of relationships between predators and their prey for species of the Knysa estuary; (b) the derived consumer overlap graph for predators 12, 13, 14 and 15. (From Sugihara, 1984.)

Sugihara concludes that stable food webs must show rigid circuitry in the disposition of consumer overlap relationships, such that graphical models of overlap, as in Figure 3.2, shall in themselves show physical and mathematical rigidity.

Rigid circuit graphs, required for food web stability, are characterized by a structure in which every circuitous path is shortened by a chord. (Such graphs are also often referred to as 'triangulated' since all generated subgraphs contain no more than triangular circuits.) Appreciation of the nature of rigid circuitry is perhaps best appreciated using a physical model analogue. Pressure applied to the top left hand corner of outline (a) in Figure 3.3 will distort the shape of the model from a square to a rhomboid; insertion of a 'cross-strut' (b) will enable the model to resist such distortion. Model (c) is likewise 'strengthened' by insertion of a cross-member to give model (d).

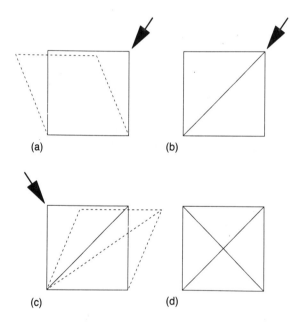

Figure 3.3 Rigid circuit models. Pressure applied to the top right hand corner of model (a) will distort the shape from a square to a rhomboid; insertion of a cross-strut (b) will enable the model to resist such distortion, but the model is still distorted by pressure from the opposite corner. In (d) the model finally displays rigid circuitry, with any three points within the matrix directly connected.

Sugihara presents an array of rigid circuit and non-rigid circuit graphs as in Figure 3.4. Applied to the real example of resource use patterns

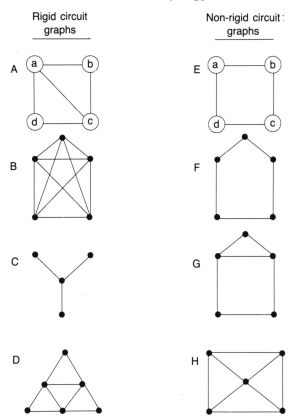

Figure 3.4 Examples of rigid and non-rigid circuit graphs. (From Sugihara, 1984.)

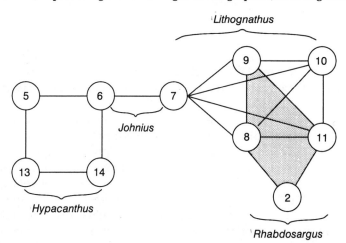

Figure 3.5 The food web of the Knysa estuary is based upon rigid circuits. (From Sugihara, 1984.)

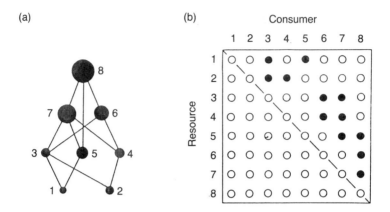

Figure 3.6 Upper triangularity in food web design. (a) A simple eight-species food web to illustrate a trophic hierarchy produced under the assumption that species feed only on others smaller than themselves. (b) The web can be represented as a matrix in which feeding links are shown by closed circles. Species are ranked by body size along rows and columns from smallest to largest; all feeding links are seen to lie above the leading diagonal and the matrix is thus 'upper triangular'. (From Lawton and Warren, 1988.)

and overlap between four fish species in the Knysa estuary (Figure 3.2), Sugihara derives a model for web linkages as in Figure 3.5 which does indeed display such rigid circuitry. (Extensions to such analysis are presented by Pimm, Lawton and Cohen, 1991.)

Perhaps a somewhat simpler topological constraint is recorded by Cohen et al. (1985, 1986) or Warren and Lawton (1987) in the observation of universal upper triangularity in food webs. Upper triangularity, as we have noted, implies strict transitivity in feeding relationships within a web, so that there are no loops of the kind where species A eats species B eats species C eats species A; or species A eats species B eats species A. Figure 3.6 (after Lawton and Warren, 1988) presents consumer relationships among members of an eight-species web as a two way matrix. All recorded consumer links are displayed in dark circles and such links may be seen to be restricted to the upper triangle of that matrix.

Food web topology

The observed upper triangularity of webs is of crucial importance: many of the other properties of stable food webs which have been characterized – such as constant ratios of predators and prey within a web or limits to the number of links in which any one organism may be involved within a web – follow automatically if food web relationships are upper triangular (Cohen and Newman, 1985). Thus, just as earlier we considered alternative explanations for the observation that food chains are in general rather short (explanations based on considerations of energy flows or unstable population dynamics within complex webs), we find that while we have thus far proffered dynamical explanations for the various structural properties observed in food webs, many of those same properties follow of necessity (Table 3.2) if webs are arranged in a transitive and hierarchical fashion (i.e. are upper triangular). If some simple explanation for upper triangularity in real-world webs may be found, then more complex dynamical explanations for many other web properties are not required (Lawton and Warren, 1988; see also, Pimm, Lawton and Cohen, 1991). Such explanation could be provided simply by the observation that in general a hierarchy of body size obtains; predators are in general larger than their prey, parasites smaller than their hosts. A simple transitive hierarchy of size like this would be enough to impose upper triangularity in real world webs.

Which explanation should we accept? Lawton and Warren advocate a middle course. They note that although dynamical models based on simple Lotka–Volterra mathematics provide a straightforward, unified explanation for structure in food webs, the worry remains that they may sometimes give the right result for the wrong reasons. The predictions they make are very sensitive to initial assumptions; (Lawton and Warren note again for example, that models emphasizing 'donor-control' dynamics, rather than 'top-down' control (pages 41–2 here) lead to very different conclusions about web structure.)

They note in addition that an unknown, but certainly not trivial, proportion of the links identified in real-world webs have insignificant dynamical consequences for consumer and consumed, and comment that for such relationships it makes little sense to invoke Lotka–Volterra dynamics. However, many of the links within such webs undoubtedly do involve strong interaction; for this subset, Lotka–Volterra models may indeed be more appropriate and may be making the right predictions for the right reasons. Lawton and Warren conclude that if the real world is characterized by partially-compartmented systems where small groups of strongly interacting species are only loosely connected to other species groups within the overall matrix, themselves highly interactive, then perhaps both dynamic and static processes in harness yield the patterns observed.

Whatever the reality, the recognition that both static and dynamic approaches to food web analysis offer such convergent conclusions as to what structural patterns might be expected within community webs,

Table 3.2 A summary of patterns found in real and stable model food webs – and their explanation by dynamic or static models of food web assembly (From Lawton and Warren, 1988.)

Observed Patterns	Theoretical Explanations	
	Standard dynamic models in Lotka–Volterra form	Cascade model generated by body-size (no dynamics)
Feeding loops absent.	√	√
Links per species constant. Connectance declines hyperbolically with increasing number of species.	√	Assumed by model
Constant proportion of basal intermediate and top species. (Constant predator/prey ratio.)	√	√
Constant proportion of links between basal, intermediate and top species.		√
Omnivory rare, except in insect–parasitoid and donor controlled webs.	√	(X)
Webs from more constant environments have more connectance, more variation and more omnivory.	√	
Food chains are short.	√	√
Food chains are shorter in two dimensional habitats.		
Webs are not compartmented except at habitat boundaries.	√	
Food webs are interval.	X	(√)

Several of these patterns can be explained either by models based on Lotka–Volterra dynamics or by the static cascade model in which body size may play a crucial role. Alternative theoretical explanations are also available for a number of these patterns, but are not discussed in detail here. A √ signifies that theory predicts the observed pattern; a X signifies that it does not. A blank cell denotes that no attempt has been made to link that combination of pattern and theory. Parentheses signify more tentative or uncertain predictions.

assures real confidence in those conclusions – and at the same time restores to an extent, one's confidence in the dynamic models themselves.

The fact that – despite their emphasis on top-down interactions – they give very similar predictions about structural constraints in web design to those reached independently from Cohen and Newman's Cascade models suggests that the majority of their predictions are extraordinarily robust and not as sensitive to the models' assumptions as might have been feared. And the fact that two entirely independent modelling approaches reach very much the same conclusions about the necessary structure of food webs perhaps instils confidence in the certainty of the conclusions, whatever the cause!

4
Niche theory: niche packing and community structure

Sugihara's analyses of topological constraints on food web structure, with which we ended the last chapter, focused in large part on considerations of overlap in patterns of resource use among web members (Sugihara, 1984). Studies of overlap in resource use, and of the way in which the available resources are partitioned amongst the different members of a community, in themselves offer a rather different perspective on the way in which interaction within a community may impose structure: opening a window on how the community's members may be organized, or 'packed', within resource space.

In such a context, any organism's 'position' within a community may be defined in terms of its pattern of utilization of a range of resources – and its interaction with other organisms over those same resources. This abstraction of the organism's position within the community, and resource relationships with others around it, is expressed in the ecological concept of the **niche**. First coined as a descriptive term to distinguish between the simple 'address' of any organism in the natural world (the types of assemblages in which it might be found) and its 'profession' (the role the organism plays within its community; Elton, 1927), geometric characterization of this hitherto rather nebulous concept (Hutchinson, 1957, 1959) made it at last amenable to more formal analysis. Characterization of the niche as the fitness of an organism or a population in a multidimensional environmental space (Levins, 1968) reflected as the envelope of resource utilization functions on each separate dimension in turn, opened the way for a series of theoretical analyses of resource relationships within communities, and subsequent empirical studies attempting to attach real values to the various niche parameters defined. The fundamental driving force behind both theoretical and empirical analyses was the presumption that an understanding of the factors determining niche structure, factors determining or constraining relationships between adjacent niches, might offer significant new insights into further structural constraints on the way in which communities are organized.

4.1 THE THEORY: A RECAPITULATION

Along any given resource axis the niche occupied by an organism may be considered in terms of an expressed pattern of resource use; its niche and those of others around it may be reflected on each resource axis in terms of some simple utilization function, portraying the range of resources exploited along some continuum and the extent of use of those resources (Figure 4.1). This utilization function may be defined mathematically for each individual or population in terms of its range, midpoint (mode or mean), shape and breadth.

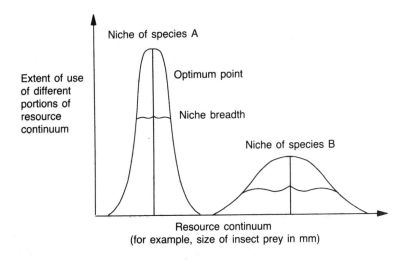

Figure 4.1 Diagrammatic representation of an organism's niche as the pattern of resource utilization along a given single resource dimension. (See text for explanation.)

In effect, purely physical limitations will determine the outer limits to the organism's occupied position in resource space. Because of morphological limitations such as mouthpart size or body weight, or physiological constraints such as enzyme systems of limited temperature tolerance, it simply cannot operate outside some restricted range of conditions. Such mechanical and/or physiological limitations (commonly considered as the organism's physiological 'tolerance limits'; Shelford, 1913) will determine for each organism a **fundamental niche**; a limited sector within the resource continuum overall, outside which parameter values are beyond the limits of its operational range. Such resource utilization curves are characteristically bell-shaped normal distributions. Any given individual will have an optimal point along the resource continuum where its performance is maximal (for example, a particular optimal size of insect prey determined by an individual bird's beak size). Its foraging will clearly not be restricted exclusively to this optimum size of insect; rather it will feed over a range of prey items slightly larger and slightly smaller than this optimum; nonetheless, its efficiency of capture/handling of prey items will decline with insect sizes further and further away from the optimum – and in consequence its overall pattern of resource use will tend to be normally distributed around the optimum point. By the same token, within a population of organisms, variation in morphological characters tends to be normally distributed around some midpoint; beak sizes of insectivorous birds will vary normally around some mean – and the expressed resource utilization curve of the entire population will once again show normal distribution.

In practice, organisms rarely occupy their entire fundamental niche; biotic interactions such as competition, predation and other local factors such as the predictability and availability of resources combine to shape a **realized niche** as a subset of the fundamental or 'pre-interactive' niche (terms after Hutchinson, 1959; Vandermeer, 1972). The resource curve defining this realized niche is no longer necessarily normal (Roughgarden, 1974; Southwood, 1978). The actual form of the realized utilization curve will be affected by patterns of resource availability; the position (midpoint), the breadth (variance) and the total range of resource values over which the exploitation function is spread will be influenced by competition, predation, and once again, the availability and predictability of resources.

Thus any individual or population may be excluded from part of the entire range of resources theoretically available to it within its fundamental niche, because that part of the resource continuum is pre-empted by some strong competitor (Figure 4.2), or because by avoiding that segment of the resource continuum (now perhaps temperature regime or habitat) it escapes overlap with some potential predator. Interspecific competition or predation, if equally balanced on either side of the fundamental niche, may cause an even 'withdrawal' at both extremes of the potential niche; if effective only at one end or the other of the potential range, such interaction may produce a skew in the realized niche. The overall shape of the resource utilization function may be influenced by such kurtosis, but may also be directly influenced by the pattern of resource availability. If resources are not equally available across the whole range encompassed by the realized niche, resource use patterns may show deviations from

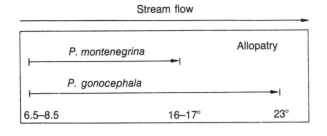
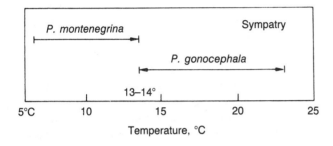
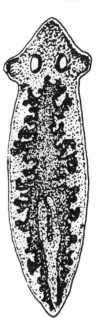

Figure 4.2 Biotic interactions shape the realized niche. The Figure shows the distributions of two freshwater platyhelminth flatworms, *Planaria montenegrina* and *P. gonocephala* along temperature gradients. Each species is restricted to a smaller range of thermal conditions when they occur in the presence of the other (below) than when in allopatry (above). (After Beauchamp and Ullyett, 1932.)

the simple bell-shaped curve expected, purely through lack of availability of resources over some range within the spectrum.

The overall **breadth** of the expressed niche – an indicator of how much of a generalist or specialist is the individual or population concerned – is itself affected by competition, predation and predictability of resources. Through their earlier effects of excluding the organism from use of certain parts of the available resource continuum, interspecific competition and predation may already have had an influence on overall niche breadth, causing patterns of resource use to contract away from areas of the resource continuum where the organism would be exposed to strong negative interaction. But each individual within a population is also exposed to continuous competition from its own conspecifics; if competition with other species is not severe, **intra**specific competition will force individuals to try to exploit those parts of the potential resource range where competition with their fellows is reduced. While at the level of the individual organism the effects of predation and of interspecific and intraspecific competition are the same, exerting pressure towards an overall reduction in expressed niche width, at the level of the population intraspecific competition – by encouraging each individual within the population to specialize in areas of the resource continuum less fully exploited by other conspecifics, will promote diversification of individuals within the population and will act to broaden the pattern of resource utilization of the population as a whole. Intraspecific competition, while it causes a reduction of individual niche width, will result by converse in a broadening of the population niche.

Finally, the pattern of resource use ultimately expressed on any one resource axis – and specifically this last parameter of niche breadth – will be influenced in addition by predictability of resources. While the powerful forces of predation and interspecific competition impose a tremendous pressure on any organism to restrict the breadth of resource use to that point where negative interactions with all others are minimized, no organism can afford to become too much of a specialist in an environment where conditions and the availability of resources may be unpredictable and changeable; here is a clear case for remaining something of a generalist and ensuring that the realized niche embraces sufficient diversity of resources to 'weather' unpredictable changes in availability, to enable the individual or population to continue to function even at the nadir of resource availability.

4.2 THE EVIDENCE? NICHE SHIFTS AND CHARACTER DISPLACEMENT

Despite the intuitive simplicity of niche theory, many of its anticipated consequences are harder to establish in practice. Even some of the fundamental assumptions remain unproven, and field evidence for the expected patterns of resource use among co-occurring species sets is far from unambiguous.

One of the most obvious implications of niche theory as rehearsed in the last few paragraphs is that we should be able to observe a change in resource utilization patterns of any organism in the presence of strong competition; the loss of fitness experienced due to competition in the region of niche overlap would be expected to lead to changes in ecology of resource use of one or both species involved, both in the short term (through behavioural modification of resource use patterns) and in the longer term, through evolutionary adaptation, if such shifts in resource use are maintained. We cited, in simple illustration of such a response, Beauchamp and Ullyett's observations on temperature range of *Planaria gonocephala* when sympatric with the congeneric *P. montenegrina* (Beauchamp and Ullyett, 1932; Figure 4.2 here). The observations by Huey, Pianka, Egan and Coons (1974) of quantitative shifts in the size of termite prey taken by *Typhlosaurus* skinks in the Kalahari desert seem equally persuasive. Huey et al. compared both feeding ecology and morphology of two adjacent populations of *Typhlosaurus lineatus*, one sympatric with the slightly smaller *Typhlosaurus gariepensis*, one in allopatry. Both subterranean skink species feed exclusively on termites, largely on the same species; in sympatry with *T. gariepensis*, *T. lineatus* concentrates far more on the larger castes and larger species within the prey spectrum. In concert with such change of dietary focus, Huey et al. discovered that overall body size of individual *T. lineatus* increased abruptly at the boundary of its sympatry with *T. gariepensis* – head size, too, increased disproportionately in sympatry.

Indeed, there are any number of examples in the literature where the resource use pattern of a species differs in some way in the presence of other organisms from the pattern of use expressed when the same species occurs in isolation; observations of such differences are commonly accompanied by measured change in some morphological or physiological characteristic which appears to reflect an adaptation to such a maintained shift in the pattern of resource use (the **character displacement** of Brown and Wilson, 1956). A wealth of illustrations of this kind may be found in almost any introductory text, together with the broader observation that there seems to be in general some limiting similarity in body size between sympatric congeners. This more general observation (Hutchinson 1959; and for example Schoener, 1965; Grant, 1968; Fenchel, 1975; Pulliam, 1975) suggests, again, a minimum required distance between two species in resource space and once more argues powerfully for competition as the causal factor (see also Greene, 1987).

But despite the intuitive appeal of such examples and their apparent cogency, our conclusions are still in practice only deductive. Except in rare experimental cases (e.g. Crowell and Pimm, 1976) or others with well-established recent histories (Diamond et al., 1989) we cannot prove that shifts in resource use, character displacements or regularity of separation of congeners in ecomorphological space are the result of competition; as noted on page 34, we are merely presented with a *fait accompli*.

4.3 RELATIONSHIPS IN RESOURCE SPACE: NICHE OVERLAP, NICHE SEPARATION

Further, not only may we not prove that such discontinuities in resource use as we can enumerate necessarily reflect an impact of interspecific competition in shaping the realized niche in natural communities; in addition, we commonly encounter instances of very substantial overlap in the patterns of resource use expressed. A necessary consequence of classical niche theory as expounded above should be that, by whatever mechanism it is accomplished, overlap between adjacent species in resource space should be minimized; each niche will be clearly separated

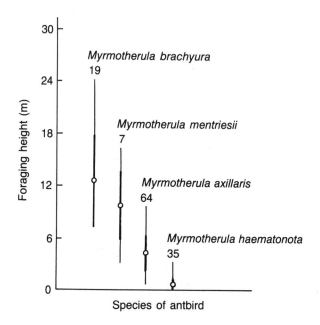

4.3 Foraging heights of four sympatric species of antbirds, *Myrmotherula*. Mean foraging heights are represented by open circles, standard deviations by thickened vertical bars. (After MacArthur, 1972, from data of Terborgh.)

from others in resource space in the phenomenon of **ecological separation**. Schoener (1974) indeed has suggested that the limits to similarity of resource utilization among coexisting sets of species should result, in theory, in rather regular spacing of species in resource space. A number of sets of species that do differ from each other primarily along one single dimension of the niche do indeed appear to be separated by rather constant amounts (e.g. Orians and Horn, 1969; MacArthur, 1972); a particularly clear – and oft-quoted – example of this is derived from Terborgh's observations on foraging heights of various sympatric species of antbirds, *Myrmotherula* (Figure 4.3). Nothing in ecology is as clear-cut as we might wish, however, and despite the numerous examples one may quote of clearly-defined separation in resource use among consumers, there are equally numerous cases where we may encounter considerable overlap of niches. Indeed quite extensive overlap in resource use seems more the rule than the exception.

In some cases such overlap is easily explained. Where resources are superabundant, for example, substantial overlap may be tolerated without strong competition; competition for any resource (as the motive force for niche separation) will only occur when resources are actually or potentially limited in supply. By definition, if there is more than enough for everyone, they cannot be competing for it; if resources are genuinely present in superabundance, avoidance of niche overlap is unnecessary. In such cases, if resources become restricted once more, clear separation between the community members may be re-established as each withdraws from the zone of overlap to a characteristic 'refuge'. Clear illustrations of such tolerance of overlap in a context of superabundant resources are provided by Nilsson's (1969) studies of prey taken by several species of diving ducks wintering together off the coast of Sweden and Reynoldson and Davies' classic analysis of dietary overlap among four lake-dwelling flatworms, *Polycelis tenuis, P. nigra, Dugesia polychroa* and *Dendrocoelum lacteum* (Reynoldson and Davies, 1970). In both cases, later restriction of resources resulted in reduction of overlap, as each species re-emphasised the proportion of its diet derived from its own specific 'food refuge'. Amongst the lake-dwelling triclads studied by Reynoldson and Davies, while the food niches of *Polycelis, Dugesia* and *Dendrocoelum* overlap, *Polycelis* species feed to a greater extent than do the others on oligochaetes, *Dugesia* on gastropods and *Dendrocoelum* on *Asellus*; within the oligochaete refuge, *P. tenuis* has a specific refuge in Tubificidae, *P. nigra* in Naididae.

Alternatively, high overlap along one resource axis may be accommodated by separation along some other dimension of the niche; the apparent overlap is an artefact of our original resolution of the multidimensional niche into a series of single dimensions. Thus two species, both earthworm specialists feeding exclusively on earthworms, and hence showing complete overlap in regard to diet, may in effect show no real overlap in multidimensional resource space because one feeds entirely in pasture, while the other forages only in woodland. More formally, we may (after Pianka, 1976, 1981) translate the resource utilization patterns of a set of species on each of two separate resource axes into a two-dimensional projection (Figure 4.4); while on axis 1, the resource utilization functions of species B and G are totally enclosed one within the other, clear separation between B and G in their use of some second resource results in effective separation in two-dimensional space.

Despite such rationalization, substantial residual niche overlap is often apparent among sympatric species, whether measured along one resource axis, or within multidimensional space. Even during the season of most intense competition, when highest separation in resource use is expressed, the extent of dietary difference between the four species of Reynoldson and Davies' flatworm assemblage averaged less than 46.2% of total food consumed – and there are many further examples where organisms appear to co-occur in a potentially competitive situation with

Niche theory

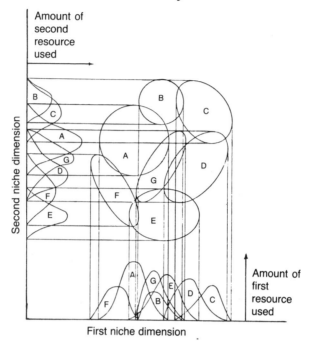

Figure 4.4 Projected niche relationships of a number of species in two resource dimensions, showing that pairs of species with substantial overlap along one dimension can avoid competition by niche separation along another dimension. (From Pianka, 1976.)

even greater overlap between their niches. Consideration of the degree of separation along other resource axes, and derivation of some multi-dimensional measure of overlap may ease the problem, but does not banish it completely. Conventionally, estimates of niche parameters (breadth, overlap) along each separate dimension may simply be multiplied together to produce an estimate of relationships in multidimensional space, if resource dimensions considered are effectively independent of one another (Pianka, 1975; May, 1975a). If resource dimensions are not entirely independent, such calculation leads to a slight overestimation of the actual degree of overlap between two niches, while if niche dimensions are totally interrelated, a more accurate composite derives as the arithmetic mean of the separate individual estimates (May, 1975a).

Using such a theory, Leuthold has attempted to quantify the degree of overlap experienced between four species of browsing ungulates in the Tsavo National Park in Kenya (Leuthold, 1978), calculating niche overlap along each of three dimensions (food plant type, browsing level, habitat preference) as:

Relationships in resource space

$$\alpha_{j,k} = \frac{\Sigma p_{ij} p_{ik}}{(\Sigma p_{ij}^2 \Sigma p_{ik}^2)^{1/2}} \quad \text{after Pianka, (1975)}$$

where $\alpha_{j,k}$ represents the degree of mutual overlap between the j-th and k-th species and p_{ij}, p_{ik} are the proportions of the i-th resource utilized by j and k respectively. (The index can assume values between zero and unity.)

Leuthold has shown that there is substantial overlap between the various species pairs in each separate dimension; while calculation of the combined overlap for the three dimensions together restores some degree of separation, a residual overlap is still apparent at a level as high as 0.54 (Table 4.1).

Table 4.1 Ecological separation among Browsing Ungulates in Tsavo National Park, Kenya

Species pair	Season	Niche dimension			Combined overlap
		Food	Habitat	Browsing level	
Kudu/gerenuk	Green	0.97	–	1.0	–
	Dry	0.99	0.56	1.0	0.55
Kudu/giraffe	Green	0.98	–	0.67	–
	Dry	0.98	0.97	0.32	0.31
Gerenuk/giraffe	Green	0.95	0.68	0.67	0.43
	Dry	0.99	0.50	0.32	0.16
Kudu/rhino	Green	0.50	–	0.78	–
	Dry	0.61	0.93	0.86	0.49
Gerenuk/rhino	Green	0.38	–	0.78	–
	Dry	0.48	0.69	0.86	0.28
Giraffe/rhino	Green	0.39	–	0.21	–
	Dry	0.47	0.89	0.16	0.07

The table shows the indices of overlap calculated for each of three resource dimensions and the resultant combined overlap calculated for each pair of species in the green and dry seasons of the year, using data from Leuthold (1978). Combined overlap is calculated as the product of the indices in each separate resource dimension. (After Leuthold, 1978.)

Similar calculations are presented by Putman (1986) for overlap between the various large herbivores in a multi-species assemblage in the New Forest in Hampshire. Overlap indices are calculated in the same way for sympatric populations of fallow deer (*Dama dama*), sika deer (*Cervus nippon*), roe deer (*Capreolus capreolus*) and free-ranging populations of domestic cattle and ponies. Table 4.2 shows the degree of niche overlap experienced between each species pair in habitat use and diet; calculation of multidimensional overlap reduces these values, but a substantial residual overlap is still apparent. While other evidence suggests that cattle and horses are true competitors, resolved multidimensional overlap between other species pairs remains as high as in Leuthold's analyses, at 0.55 (Putman, 1986).

Table 4.2 Resource use overlap and separation among large herbivores in the New Forest, southern England

		Cattle	Ponies	Fallow	Sika	Roe
SPRING (February – April)	Cattle	*				
	Ponies	0.63	*			
	Fallow	0.48	0.28	*		
	Sika	0.40	0.20	0.49	*	
	Roe	(0.20)	(0.14)	(0.39)	(0.53)	*
		Cattle	Ponies	Fallow	Sika	Roe
SUMMER (May – July)	Cattle	*				
	Ponies	0.70	*			
	Fallow	0.05	0.12	*		
	Sika	0.09	0.12	0.55	*	
	Roe	(0.14)	(0.14)	(0.35)	(0.32)	*
		Cattle	Ponies	Fallow	Sika	Roe
AUTUMN (August – October)	Cattle	*				
	Ponies	0.58	*			
	Fallow	0.16	0.26	*		
	Sika	0.14	0.23	0.55	*	
	Roe	(0.20)	(0.17)	(0.45)	(0.31)	*
		Cattle	Ponies	Fallow	Sika	Roe
WINTER (November – January)	Cattle	*				
	Ponies	0.42	*			
	Fallow	0.03	0.12	*		
	Sika	0.02	0.21	0.43	*	
	Roe	(0.16)	(0.14)	(0.68)	(0.37)	*
		Cattle	Ponies	Fallow	Sika	Roe
ALL MONTHS COMBINED	Cattle	*				
	Ponies	0.70	*			
	Fallow	0.17	0.22	*		
	Sika	0.17	0.20	0.53	*	
	Roe	(0.18)	(0.15)	(0.43)	(0.37)	*

$$\text{Overlap} = \frac{\Sigma p_{ia} p_{ja}}{[(\Sigma p_{ia}^2)(\Sigma p_{ia}^2)]^{1/2}}$$

The table shows calculated combined overlap between each pairwise combination of species (fallow, roe and sika deer, domestic cattle and ponies) in each season of the year, based upon assessed overlap in patterns of habitat use and diet. (From Putman, 1986.)

4.4 LIMITS TO OVERLAP

Given illustrations such as these, just what are the limits to overlap in resource use between sympatric species? Can we derive any theoretical predictions of what should be the 'limiting similarity of coexisting species'? (MacArthur and Levins, 1967.)

MacArthur and Levins examined the relative efficiency of exploitation of a set of resources by two comparative specialists whose resource use spectra along some linear axis were more, or less, distinct and a single generalist species (Jack of all trades) whose maximal efficiency lay somewhere between that of the two specialists. Efficiency of exploitation of the generalist and each of the two specialists was assessed on a fine-grained mixture, in equal proportion, of the preferred resources of each of the two specialists; the more similar the resource utilization functions of each specialist in isolation, the more likely a generalist, with intermediate resource utilization would be able to outcompete either or both specialists. In effect a single generalist becomes more efficient at exploiting the mixed resource than either specialist when the resources favoured by each of the two specialists are so close that their independent resource use curves overlap within two standard deviations of their individual optima. This point, where a single generalist may then exploit the double niche more effectively than either specialist, might be considered the point beyond which niche overlap between the specialists becomes intolerable – and thus the limit to overlap between adjacent species along any one resource axis (MacArthur and Levins, 1964).

This analysis is clearly oversimplified: considering efficiency of exploitation of unlimited resources in a mixture of precisely equal proportion. In a later consideration of the same issue, MacArthur and Levins (1967) assessed the limiting similarity/overlap between species due to competitive exclusion when resources might be presumed to be limiting, calculating a limiting similarity between the niches of 0.54. More recent models have refined this analysis still further (May and MacArthur, 1972; May 1975a; Fenchel and Christiansen, 1976). Taking just one of these studies by way of illustration, May and MacArthur's analysis suggests that the limits to niche overlap along any one resource dimension are such that the distance between the midpoints of adjacent resource use curves must exceed one standard deviation ($d>w$). In the same notation, the conclusions of MacArthur and Levin's earlier deliberations – leading to a presumption that overlap should not occur within two standard deviations of the midpoint – may be expressed as $d>4w$; minimum distance of the point of overlap of the resource utilization curves should be two standard deviations from the midpoints of each. Such contrast clearly demonstrates that the results of any investigation seeking limits to similarity depend heavily upon the models used and on initial premises;

further, predictions derived vary over a very substantial range. It is perhaps a whimsical coincidence that the calculated maximum overlap of 0.54 of MacArthur and Levins' (1967) analysis, compares so neatly with the residual multidimensional overlap resolved in the studies of Leuthold (1978) or Putman (1986).

Whatever the maximum possible overlap in theory, in real communities overlap will in any case rarely extend to this theoretical potential. Real overlap will always be less than that theoretically possible as a result of the effects of 'diffuse competition' within the community (MacArthur, 1972). Even if the effects of direct competition between major protagonists are taken into account in calculation of overlap, effects of a variety of indirect interactions from other members of the community (diffuse competition, competitive mutualism, etc.) may still have a further, unpredictable, effect on permissible overlap. Maximum permissible overlap as calculated here should in theory remain a constant; maximum tolerated overlap in the real community has been shown to depend heavily on the number of species in the community and the pattern of species packing (Pianka, 1973; Pianka, Huey and Lawlor, 1979).

4.5 IMPLICATIONS OF NICHE DYNAMICS FOR COMMUNITY STRUCTURE: NICHE PACKING

The conceptual simplicity of niche theory and the relative ease with which niche relationships could be modelled and explored, together with the traditionally central position of the niche concept in community ecology, led perhaps to somewhat unrealistic expectations for the theory in the resolution of process and structure in real communities. While various attempts have been made to determine some of the rules governing internal structuring of species assemblies from analyses of separation or overlap in real communities, such efforts to apply our developing understanding of niche dynamics to an analysis of niche-packing within communities have met with only limited success.

It is in fact frustratingly difficult to synthesize the conclusions of this work, for the literature itself is confused and conflicting. Different authors appear to work from different premises – and have worked in markedly different communities or sub-communities, which may well not obey the same 'rules'. Thus, for example, in his analysis of changes in niche relationships with increasing species diversity in assemblages of desert lizards, Pianka (1973) found evidence of decreasing niche overlap with increasing species number. Cody (1974), using methods essentially similar to those of Pianka, in an analysis of the structure of shrubland bird assemblies came to the diametrically opposite conclusion: finding that average overlap between adjacent niches within the community

Implications for community structure

increased with increasing diversity. In addition, 'packing' is variously defined by different authors – in terms of

1. the number of species that can be accommodated per unit volume of resource (or, resolved to a single dimension, per unit distance of a resource continuum) (e.g. Roughgarden, 1974; Rappoldt and Hogeweg, 1980);
2. the closeness of packing of resource utilization distributions for non-overlapping species (Pianka, 1975);
3. the extent of overlap observed in resource utilization within the community, or between adjacent pairs of species within the community (MacArthur, 1970; Roughgarden and Feldman, 1975).

While all these do indeed represent facets of the way species are packed into a community, all represent rather different facets. The number of species which can be accommodated is a measure of community fill; the pattern in which those species are arranged within niche space (their actual spatial relationship to one another) is an aspect of packing design; the closeness of packing of overlapping or non-overlapping species, or the degree of overlap expressed are all derived functions of the extent of community-fill, packing design and the types of organism involved; nor are these separate elements necessarily related to each other in any simple way (Putman and Wratten, 1984). Thus extent of resource overlap, for example (the measure most commonly referred to as 'species packing'; e.g. Roughgarden and Feldman, 1975) is not, as has sometimes been implied, a valid index of community fill; indeed it is not necessarily related to community fill in any way. While it is true that increased 'fill' of a community may well be reflected in an increase in niche overlap in many instances (e.g. Cody, 1974), in other circumstances an increase in the number of species within a given resource space may equally be accommodated without such increase in overlap, but through restriction of average niche breadth (e.g. Thorman 1982), or by expansion of the actual resource range exploited by the combined assembly (as in Pianka's (1973) study of desert lizard assemblies where increased species number was accompanied by decreased niche overlap). Such contradictory results, taken together with a lack of certainty as to precisely what should be the theoretical limit to permitted overlap, have left such lines of enquiry disappointingly unfulfilled.

Even were our understanding of niche structure and niche relationships within a given assemblage of species crystal clear, it is apparent that the application of such understanding to determine rules of community structure via constraints on species packing is beset with difficulties.

The difficulty of deciding upon appropriate measures in field studies, and uncertainty in the interpretation of those measurements once derived, have resulted in many expressions of increasing doubt about the validity of existing indices of niche breadth, niche similarity and overlap (e.g. Colwell and Futuyma, 1971; Hanski, 1978; Smith and Zaret, 1982) and of the actual interpretation of data once assembled (Rusterholz, 1981; Holt, 1987). Such doubts have always been respectfully presented, 'sharing' uneasiness in the interpretation of field data in relation to underlying theory. In this concluding section I will take the stronger stance, arguing that empirical studies of niche relationships in the field, despite the elegance of their theoretical origins cannot reveal anything of the incidence of competition within the community nor of any other process imposing structure upon the community. I suggest that, in effect, the concept and measurement of the niche is useful as a simple descriptor – in defining for a given individual or population its position within the community – but has and can have no more general predictive value.

4.6 INITIAL DOUBTS: PROBLEMS OF APPLICATION AND INTERPRETATION

Perhaps the most fundamental problem with such analysis lies in the difficulty of accurate measurement of true overlap in field studies in the first place and in interpretation of that overlap. If interrelationships between the niches of organisms within a community are going to affect species packing it is clear that they must do so through the medium of competition, actual or potential. Yet even the inference of the existence of competition from considerations of niche overlap is far from straightforward. Many general reservations have been voiced over the years. The most commonplace arguments were neatly summarised by Rusterholz (1981) (following Colwell and Futuyma (1971) among others): 'Competition can be inferred from niche overlap only if it is known that resources are in limited supply. The common use of non-limiting resources by different species should not be indicative of competition.'

Areas of overlap between the resource utilization curves of adjacent species reflect a potential for interaction, and indeed some degree of niche overlap is clearly a necessary prerequisite for interspecific competition. However, interpretation of such overlap in terms of expressed competition is fraught with difficulty. Competition will only result from overlap in resource use if those resources are themselves actually or potentially limited in supply (page 67), something often very difficult to establish in field studies. Thus, even when a relatively high degree of overlap can be demonstrated in empirical studies, interpretation in terms of competition must be cautious. High observed overlap could indeed imply competition where resources are limited, but it could equally well be argued that such overlap must imply a lack of competition (since if competition were being experienced some degree of shift in resource use might be expected). Further, as already observed, high overlap in use of

Additional problems

one resource might well be accommodated by separation along some other dimension of the niche. In any analysis of niche overlap, therefore, we should properly consider the full multidimensional projection.

In practice few authors (e.g. Schoener, 1974; Hurlbert, 1978) have incorporated resource availability into measures of niche overlap or competition, and resolution of multidimensional overlap is itself rather contentious. Here again we may encounter problems of both calculation and interpretation. Conventionally multidimensional niche overlap may be computed as the product of measured undimensional overlap indices provided the resources considered are independent, or as the arithmetic mean of the separate overlap indices if the resources considered are dependent upon each other (Pianka, 1975; May, 1975a). In such a case our hypothetical earthworm specialists of page 67 show total overlap in diet ($\alpha = 1.0$) but total separation in habitat use ($\alpha = 0$); product $\alpha = 0$. Alternatively the six species of large ungulates within the New Forest, Hampshire, show a measurable degree of overlap in their use of forage resources and habitats (Table 4.2). Let us assume, however, that in this case differences in diet between for example, roe deer and fallow deer are a direct reflection of primary differences in habitat – one species eats more tree leaves because it spends more time in woodland. These two measures then in effect represent two different estimates of the same difference between the species. In such case average α is appropriate.

The illustrations are chosen deliberately because the resolution is clear; in many cases it is difficult to determine whether different resource axes are dependent or truly independent. Slobodkichoff and Schulz (1980) offer a device for testing independence of resource dimensions, but it is perhaps of theoretical rather than practical value, for while it may quantify the mathematical dependence between values recorded along two resource dimensions, it cannot provide information about the degree of biological interdependence of the two resource types; further, what formula should be applied in cases where the resources considered are neither fully dependent nor fully independent? In fact the problems of calculation are not too serious, in that the errors in estimation resulting from adopting the wrong procedure are non-linear; the proportional error is low when overlap is in any case high, and the error high only when overlap is in any case very low. More serious problems surface in interpretation of such multidimensional indices.

4.7 ADDITIONAL PROBLEMS: CALCULATION AND INTERPRETATION OF MULTIDIMENSIONAL INDICES

In principle, if calculated overlap in n-dimensional space is zero, competition must be avoided; after all niche overlap is a prerequisite for interaction. Yet it is relatively easy to contrive examples where conventionally calculated multidimensional overlap is indeed zero, yet competition is demonstrably still occurring. (Alternative scenarios where overlap is >0 but no competition occurs are essentially trivial.) Problems arise in relation to the type of resource considered and in the fact that there seem to be different effects on apparent niche separation or overlap of interference and exploitation elements of competition.

Holt (1987) has also drawn attention to this conundrum of severe competition in the face of a calculated niche overlap of zero; he considers two situations which might lead to such an obvious contradiction. In the first instance the problem will arise whenever different niche axes to be combined relate to totally different aspects of an organism's life history. Holt considers an example of Central American flycatchers competing fiercely for aerial insect prey, yet separated out in type of nesting site preferred. Were multidimensional overlap calculated in relation to these clearly incommensurable resource axes, it would perforce be zero. Such an example is perhaps somewhat extreme and can be resolved in this or similar cases in the recognition within gross patterns of resource use of functionally-related or functionally-unrelated subsets. Thus where pattern of habitat use, say, is to be combined with food use, data on gross patterns of habitat use overall are hardly relevant; patterns of food use should more properly be combined with habitat use while feeding. Such functionally related subsets of overall patterns of resource use may readily be recognized, and derived niche metrics calculated only within such functional sets.

Similar nonsenses, however, can also arise where qualitatively different types of axis are combined, as when temporal or spatial axes are combined with what we may call materialistic resource dimensions. Imagine two insectivorous birds – call them species C and D – which overlap heavily in terms of the insect species taken as food and the size of prey consumed (high overlap, resource dimension food). Suppose, however, that C and D are totally separated out in relation to foraging height within the canopy (overlap zero). The two resources of insect size/type and foraging height seem totally independent, thus niche overlap would be considered as product α, and equals zero. But what happens if the insects move up and down within the canopy? The same population of insects would then be predated by C and D even though they are operating at different heights. The birds therefore, while effectively separated out in terms of interference competition, still suffer severe exploitation competition. Such a problem was recognized by MacArthur himself in 1968, who wrote: 'The statement "two species coexist if their niches do not overlap" is plausibly false since by feeding in different places two species would occupy non-intersecting niches even if they both depended on and competed for the same highly mobile food supply.'

Let us take a further example. In the analysis of niche relationships

Additional problems

among the large herbivores of the New Forest, (page 69) estimates of apparent overlap were calculated in relation to diet and habitat use. But for many of the species, data are also available on temporal patterns of activity throughout the 24-hour period (Putman, 1986). Imagine two large herbivores, one primarily diurnal in habit (as are cattle), one effectively nocturnal (it might be sika deer). Let us assume that they eat the same plants, the same parts of plants, at the same browsing height (thus dietary overlap = 1), but that one uses them by day, the other by night. If time of day may be considered a resource (and in effect the separation achieved is no different from that achieved through differences in foraging height in our last example), combined overlap = 0. Yet patently, while the two ungulates may avoid interference exploitation, once more there is no question but that each will deplete the resources available to the other in prior exploitation. Time as well as space may not be a resource axis ensuring segregation.

In these examples, the resources of food and time/space are not strictly incommensurable as in Holt's earlier combination of the qualitatively distinct resources of food type and nesting requirements, but they *are* in some sense non-independent. The two resources in each case are independent in that they are not simply different measures of the same pattern of resource use (as were habitat and food use in our earlier consideration of roe deer as woodland browsers vs. fallow as grazers in open habitat; page 75). However, they are clearly interactive in some way and thus cannot simply be combined in any derivation of multi-dimensional niche overlap. Despite this, the calculation of multidimensional overlap cannot be considered totally invalid since, in the examples presented here, separation in the one case by foraging height, in the other by time of day, while it does not affect exploitative interactions between the species, does reduce interference to zero. It is clear that

1. Niche separation by certain resources relates specifically to reduction of interference; niche separation by other resources relates only to a reduction of exploitation competition; in yet other cases reduction of multidimensional overlap may reduce both components of competition.
2. The result depends on the type of resource. Habitat separation works in our earlier example of earthworm specialists characteristic of pasture or woodland habitats, because the shared prey are relatively non-mobile earthworms; it does not work where foraging height separates insectivorous birds, because the prey in that case are highly mobile relative to the degree of spatial separation.

Rusterholz (1981) offers as a major criticism of studies of niche overlap that such analyses only take into account a potential for exploitation competition, but not for interference. This is an oversimplification. What is clear is that different resource dimensions relate preferentially to one or other type of possible interaction. In our examples here it is apparent that separation by temporal or spatially related dimensions (time of day, habitat, etc.) may influence the potential for interference interactions (see also Carothers and Jaksic, 1984); separation along the more materialistic axes of food type or size, refuge character, etc. will affect exploitation competition. In relation of niche overlap to competition, resources need not be limited if the relationship considered is one of interference, while resource limitation is required if niche overlap along material dimensions is to lead to exploitation competition. Finally, combination of potentially interactive resource dimensions in calculation of multidimensional niche metrics confuses the distinct concepts of interference and exploitation interaction and results in anomalies similar to those arising from combination of more clearly incommensurable axes (Holt, 1987).

In his preliminary exploration of these problems, Holt concludes: 'These observations do not imply that overall overlap indices are valueless [merely] that one should be cautious in interpretation.' Perhaps so, but one cannot necessarily determine *a priori* whether different unidimensional resources are dependent or independent, interactive or non-interactive. Often one will be unable to recognize incommensurable resource axes (Holt's own examples are artificially clearcut). Thus any derived measure of multidimensional overlap is liable to be spurious. Further, such a measure will almost necessarily confuse resources relevant to interference interactions with those relating to exploitation, resulting not only in anomalies of zero overlap yet intense competition, but also in the loss of an important distinction as to the type of interaction involved (Carothers and Jaksic, 1984).

Whatever the complexities of calculation and interpretation of niche overlap in the field, one additional problem remains: that both theoretical and empirical studies of niche overlap consider only direct interactions between two individuals or populations. It is clear however that species pairs may also influence each other indirectly, while in addition, such pairwise interactions occur within the context of a whole community of interacting organisms. Thus the effects of diffuse competition (MacArthur, 1972; Pianka, 1976, 1981) or competitive mutualism (Levine, 1976; Vandermeer, 1980) may exaggerate or completely reverse the apparent ecological effect of one species upon another deduced from simple pairwise analysis.

Amid all these emerging problems of calculation and interpretation how useful is an abstract idea of niche overlap? Does it mean anything at all in the real world – and can it be related in any way to competition experienced and the forces structuring communities? It is arguable that in a single resource dimension, when resource use curves overlap and resources are known to be limiting, competition may be inferred (although it is equally arguable that the observation that overlap remains implies that competition is not as severe as we first imagined). In the multidimensional arena I would argue that reduction of some calculated combined overlap can tell us little or nothing about reduction of competition. I suggest that the niche concept is useful only in its original formulation as a compact descriptor of an organism's functional position within a community but has and can have no more general predictive power.

Finally, we might note that this whole card-castle of analysis presumes the individual niche as the unit of assortment in partitioning of resources. Just as in studies of food webs many authors have pointed to clear subunits within the web: 'small groups of species whose interactions are dynamically strong, embedded within a network of more trivial interactions' (Paine, 1980; Lawton and Warren, 1988), so also in analyses of resource use patterns, many studies have identified discontinuities in distribution of species in resource space, resulting in distinct clusters of strongly interacting species functionally separated by some degree from other clusters of strong interaction. Such **guilds** of strongly interacting species might prove a more appropriate unit of assortment in analyses of the mechanisms structuring the distribution of species within resource space.

5
Guilds and guild structure

One of the most persuasive illustrations of some sense of order or pattern within community structure is the recognition of an apparent regularity in the manner in which a particular set of resources may be 'apportioned' amongst a set of consumers – even in widely disparate communities. There is a striking similarity of niche-form and niche structure, of the way 'the jobs are partitioned out' in 'parallel' communities.

Thus, for example, Cody demonstrates that the division of available resources amongst sets of insectivorous birds in shrub grasslands results in the projection upon each resource dimension of almost identical niches within Californian chaparral, Chilean matorall and South African macchia (Cody, 1974, 1975). While one might imagine that the division of some continuous set of resources between potential consumers might be effected in any of an almost infinite number of ways, in practice the realized niches of each species set showed remarkable similarity in all three shrubland systems – despite the fact that each is colonized by a taxonomically distinct set of species. So close appears the match that the different species of the analogous communities even display morphological convergence in their adaptations to apparently parallel niches (Cody, 1974, 1975). Similar constancy of niche structure in parallel communities is recorded among assemblages of montane lizards in Chile and California by Fuentes (1976), for the fish assemblies of coral reefs in the Atlantic and Pacific oceans (Gladfelter, Ogden and Gladfelter, 1980) or finch 'communities' (Schluter, 1986).

That said, there is in fact considerable controversy about this apparent 'fixity of design' – and views as to how the data should be interpreted are sharply divided. Cody's conclusions have been challenged by Ricklefs and Travis (1980) and Blondel, Vuilleumier, Marcus and Terouanne (1984), while, as we have already noted, the constancy of trophic structure proposed by Heatwole and Levins (1972) within mangrove faunas has been challenged by Simberloff (1978a) who claims that even this classic

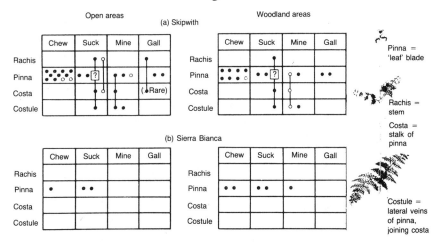

Figure 5.1 Feeding sites and feeding methods for the herbivorous insects attacking bracken (*Pteridium aquilinum*) in sites in the north of England and in New Mexico. Data (from Lawton, 1982) represent resource use in (a) open, (b) wooded sites in each case. Feeding positions of insect species exploiting more than one part of a frond are linked.

case for trophic constancy is statistically inconclusive. More powerful doubts are cast by a number of recent experimental studies more or less equivalent to those just described, but which show no such constancy. Lawton (1982, 1984), for example, found marked differences in niche structure among assemblages of herbivorous insects feeding on the above-ground parts of bracken (*Pteridium aquilinum*) in England and North America. Lawton identified a number of possible niches for occupation within the system depending on what part of the frond the insect fed upon (rachis, pinna, costa or costule) and the way in which it exploited the resource (chewing, sucking, mining, gall formation). The way in which the potential niches were filled or apportioned amongst plant colonists in the UK and North America showed no evidence of constancy (Lawton, 1982; Figure 5.1).

Note, however, that Lawton's analysis of bracken herbivores did offer comparison between the niche relationships of communities which differed dramatically in their overall species richness. 27 species of insects feed regularly upon the above-ground parts of bracken plants in Britain; only five species were found regularly on bracken in New Mexico, with a further two species recorded in Arizona. With such a small pool of potential species in Arizona and New Mexico individual local communities of bracken herbivores are also species-poor by comparison to equivalent communities in Britain; perhaps the competitive interactions which

Guilds and guild structure

might be expected to refine niche relationships within the community only begin to operate in communities approaching saturation? A more recent exploration of the niche relationships of two further phytophagous insect assemblies in Canada and the UK (using the same approach adopted by Lawton in defining potential niches and considering in each country the structure of relationships among insects feeding on red oak, *Quercus rubra*, and those feeding on aspens, *Populus tremula/tremuloides*), reaffirms striking similarity of trophic structure in each biogeographic 'pair' considered (Ashbourne and Putman, 1987; Figure 5.2). For the present at least, we should perhaps reserve judgement.

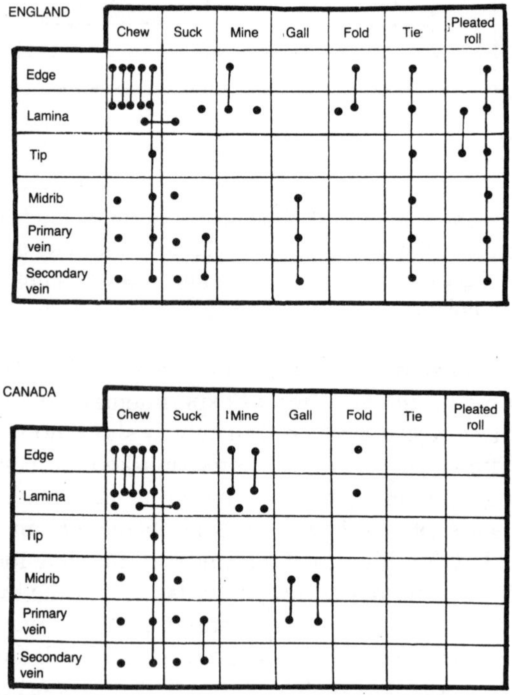

Figure 5.2 Feeding sites and feeding methods for insects feeding on red oak (*Quercus rubra*) in Canada and the UK. Data from Ashbourne and Putman (1987) are represented with the same conventions as in Figure 5.1.

Recognition of possible structure at this level, whereby available resources are partitioned amongst particular sets of species in a predictable and repeatable way, once again begs the question posed at the end of the previous chapter, of whether or not the individual niche is the appropriate unit of assortment to consider in seeking patterns of species in resource space.

5.1 EVIDENCE FOR GUILD STRUCTURE WITHIN COMMUNITIES

The idea that communities might contain functional clusters of species interacting among themselves more strongly than with other elements of the surrounding matrix is not new. We have already met such suggestion within the context of the possible recognition of structural subunits within webs of feeding interactions; subunits within trophic levels, structuring competitive relationships ('arenas of competition' (Pianka 1980)) have been mooted by Root (1967), Botkin (1974), Inger and Colwell (1977), Holmes, Bonney and Paccala (1979), and others.

Excellent reviews of the developing research into the structure and dynamics of these subclusters within the community are offered by, for example, Joern and Lawlor (1981), Adams (1985), Terborgh and Robinson (1986), and Jaksic and Medel (1990). Root (1967) first defined such **guilds** as groups of species 'that exploit the same class of environmental resources in a similar way', continuing that 'this term groups together species, without regard to taxonomic positions, that overlap significantly in their niche requirements'. It is certainly possible, on a purely subjective basis to recognize such guilds in nature: clusters of species exploiting common resources in a similar way. Thus one might identify amongst insectivorous bird assemblages, subclusters that seem to show particular similarity in their patterns of exploitation – as foliage gleaners, ground feeders, flycatching (sallying) or hawking insectivores; or amongst frugivores: ground feeders, shrub-layer frugivores or canopy frugivores (e.g. MacArthur, 1972; Holmes, Bonney and Paccala, 1979; Karr, 1980). Such guilds have been recognized among insectivorous birds (MacArthur, 1972), nectar-feeders (Feinsinger, 1976), desert lizards (Pianka, 1975, 1980), terrestrial salamanders (Hairston, 1980), but objective definition of the limits of any guild, or of its membership, proves more difficult. Further, recognition of such guilds, even if the clusters can be more objectively defined, does not necessarily imply any functional role within community dynamics. While the clusters may represent real structural subunits which focus in some way the competitive interactions within a community into specific theatres of war, the guilds identified in this way might equally arise as mere passive reflections of discontinuities in resource distribution.

In a valuable critique of the 'Abuse and misuse of the term "guild" in ecological studies', Jaksic focuses attention on these and other problems in many published analyses of guilds and their dynamics (Jaksic, 1981; see also Jaksic and Medel, 1990), noting once more that there is as yet no generally accepted protocol for objective recognition of such guilds, or for determining the limits of their membership, and that the underlying causes of apparent guild structure are unresolved. It is also apparent that despite Root's original affirmation that the term should group together ecologically related species without regard to taxonomic status, the majority of published studies of guilds and their dynamics have restricted consideration to a single taxonomic assemblage: e.g. lizards (Pianka, 1980); nectar-feeding birds (Feinsinger, 1976; Feinsinger, Swarm and Wolf, 1985); grasshoppers (Joern and Lawlor, 1981). Such focus may be readily explained, and even excused in part, by the fact that few ecologists profess expertise beyond the limits of a restricted taxonomic assemblage, but such studies in effect reduce to an analysis of resource relationship within a defined taxon rather than relationships within an entire guild of species – and may lead to false conclusions. As noted in the introduction to this book, studies of the resource relationships among aerial, vertebrate insectivores might produce mystifying patterns if bats were not considered along with birds. The classic studies by Brown and coworkers on desert granivores, embracing in their analyses rodents, passerine birds and ants (see Brown et al., 1986, for a review), or those of Feinsinger (1987) on pollinators, are among the few that truly examine relationships among all members of a complete guild.

Perhaps the problems of taxonomic tunnel vision themselves reflect in part the difficulties many workers experience in more objective delineation of true guilds within the community matrix as a whole. How does one determine the membership of a group of species 'that overlap significantly in their niche requirements'? To begin with, should one seek clusters of species along a single linear resource dimension, recognizing groups of organisms which may interact strongly in their exploitation of some specific single resource, or should guilds be defined as clusters of species with similar multidimensional niches?

5.2 FORMAL IDENTIFICATION OF GUILDS AND THEIR MEMBERSHIP

Inger and Colwell (1977) made the first explicit attempt at objective identification of guild structure within community matrices by seeking sharp discontinuities in the arrangement of resource use curves along resource axes. In effect, evidence for guild structure within the community was determined by sudden increases in variance of mean overlap computed among nearest neighbours in niche space above a given cluster size; the point at which such increase in variance is detected determines the average size of guilds within the assembly considered. The drawback with this approach, however, is that while it might identify the existence of substructure within a community matrix and provides an estimate of the average size of such clusters, it does not identify which particular species belong to which guild. Joern and Lawlor (1981) determine group membership of guilds through use of a clustering technique, progressively linking together (in unidimensional space) pairs and then clusters of species with highest overlap (Figure 5.3). In such analysis guilds are explicitly defined as clusters of species separated from all other clusters by a distance greater than the maximum distance calculated between the two most disparate members of the focal guild. Joern and Lawlor's analysis thus effectively clusters together groups of species whose competitive interactions with others are strongest within that same guild (overlap is greater with other members of the guild than with any non-member).

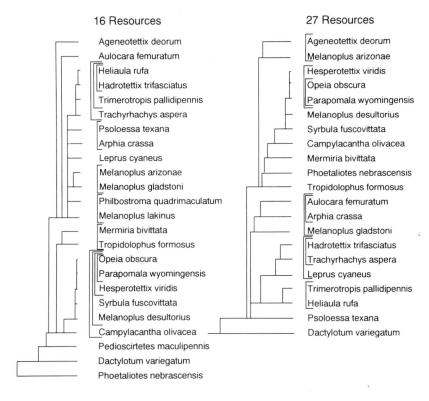

Figure 5.3 Guild structure of grasshopper assemblies in relation to patterns of habitat use. Guild structure is ascribed from cluster analysis of species in relation to expressed overlap in resource use. (From Joern and Lawlor, 1981.)

This same type of approach may be extended to analyses within multi-dimensional space using a range of multivariate clustering techniques such as Principal Components Analysis or Factor Analysis (as for example Holmes et al., 1979). The position of any species can be plotted in multivariate space with regard to its characteristic pattern of resource use along n separate dimensions. Euclidean distances may then be calculated between all the species in that space, and guilds once more defined as clusters of species separated from other clusters by linear distances greater than any within the cluster.

Both Joern and Lawlor's analysis and that of Holmes *et al.* provide a hierarchical representation of consumer species in resource space and facilitate the recognition of higher and higher levels of potential interaction. Both share the limitation however that no statistical significance can be attached to the clusters identified, relying on arbitrary thresholds or common sense to ascribe guild membership (Hawkins and MacMahon, 1989); some resolution of this is provided in the approach advocated by Jaksic and Medel (1990).

One final approach to the identification of guilds has been proposed by Adams (1985). Restricted in its application to analyses of structure along one single resource continuum it is nonetheless so different in conception and derivation from the analyses described so far that it warrants some mention here. Adams' approach is based on the so-called 'unfolding' technique first developed by Coombs (1964) for psychological analysis of behaviour preferences. The underlying logic of such unfolding analysis may be summarized as follows. If we accept that for any organism there exists along a given resource axis some optimal point where its efficiency of exploitation of that resource is maximized, but that the organism will none the less exploit, albeit to a lesser degree, a range of resources around that optimum point, then we may define that organism's preference for, or utilization of, those resources in relation to the relative magnitudes of the intervals between any given point on the resource axis and the optimum or ideal point. The ideal points of a number of species will be located on the same linear axis only if they perceive the resource dimension and so exploit resources in a similar way. It therefore follows that a guild may be defined as a set of sympatric species whose expressed preferences for a common set of key resources can be resolved to fit a single axis. The method is perhaps best understood in reference to a worked example. In Figure 5.4 data on dietary preferences of six triclad species compiled by Reynoldson and Davies (1970), Reynoldson and Sefton (1976) and Adams (1981) are represented along an axis of potential prey species. The position of the 'ideal point' for each triclad (∇) is located such that the linear distance to each potential prey type is directly proportional to their relative occurrence in the diet of that particular predator.

The six triclads comprise a guild if the prey species can be arranged in such an order along the axis, and at such linear distance from each other, that on one single axis all predators may simultaneously be allocated an ideal point which satisfies the condition that the linear distance to each prey category accurately reflects their defined preference for those prey items. In Figure 5.4, *Dendrocoelum lacteum, Bdellocephala punctata, Polycelis torva, Dugesia polychroa* and *P. nigra* may all be resolved

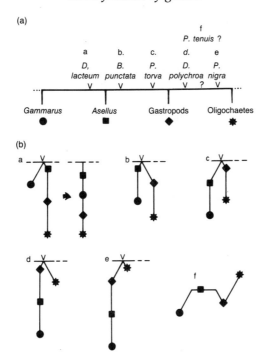

Figure 5.4 Unfolding analysis of guild structure within a group of lake-dwelling triclads. (a) The different triclad species are arranged along a one-dimensional axis of different prey types, with distance from each prey type reflecting the actual relative proportions in the diet for each predator. (b) Folding permutations required to recreate the original choice sequences of the triclads, by lifting and folding the prey sequence at the predator's ideal point V. (From Adams, 1985.)

to the same axis, while *P. tenuis* cannot be incorporated into the same guild.

Recognition, whether by intuitive or more formal analysis, of clusters of strong competitive interaction within a matrix of weaker relationships does not resolve the question of what may be the significance of such guilds. Clearly, however defined, guilds do represent clusters of species whose patterns of resource use show higher levels of average overlap than those expressed within the community as a whole: clusters where the overlap in resource use within the cluster is substantially higher than any measured overlap between guild members and non-members. To that extent they represent structural elements within the community which focus the community's competitive interactions in some way. The underlying causes of such structure, however, remain unresolved. Pianka suggests that guild structure will develop within a community as a consequence of diffuse competition (Pianka, 1980; see also Oksanen, 1987; Van Valkenburgh, 1988); others suggest that the structure observed merely reflects in some passive way gaps in the spectrum of resources available (Jaksic, 1981; Bradley and Bradley, 1985). Perhaps, however, this question of whether guilds are formed as functional clusters of species within the community or merely reflect discontinuities in resources is of more academic than practical importance. Even if in their original 'definition' within a community, guilds consist of groups of species exploiting relatively distinct packages of resources within the community, as Jaksic and Bradley and Bradley suggest, the fact remains that this will still impose some structure on competitive interactions within the community as a whole, with members of each 'guild' interacting strongly with each other, but only relatively weakly with members of other clusters.

6
Species composition and community assembly

Thus far in our discussions we have focused primarily on the interactions that occur within a community of organisms and the effects of such interactions upon community structure; we have paid less attention to the actual species composition of such communities. Yet membership of that species array is itself an important determinant of a community's structure. Returning at the same time, incidentally, to a more holistic view of the community and community processes, we will consider in this chapter those factors which may influence species composition. Factors determining diversity within the community, determining the actual number of species and their relative abundance, will be considered later in chapter 8; here we will simply consider the actual species composition. Roughgarden and Diamond (1986) define this as perhaps the central problem of community ecology as a whole: explanation of the 'limited membership' of ecological communities. In echo of their enunciation of the question we may ask: 'Why is it that what does occur together constitutes a limited subset of what might occur together?' (Roughgarden and Diamond, 1986).

6.1 SHORT-LISTING CANDIDATES: LIMITS TO TOLERANCE

At the simplest of levels, the species composition of any community is determined by two basic considerations: what species are available for inclusion within the community and what species are selected from that pool of candidates. 'Availability' of any species for possible inclusion within a given community depends in the first instance – and perhaps trivially – on biogeographic distribution (Putman, 1984), but thereafter on the abiotic properties of the environment in relation to that organism's

90 *Compositions and assembly of communities*

specific tolerance limits, and on ability to reach the community (dispersal ability).

Any organism has a restricted range of physico-chemical conditions over which it may operate. Simple curves of performance may be drawn for any organism for any particular physiological process, representing its efficiency of operation (or fitness) over a range of some physico-chemical parameter. Such curves, known as 'tolerance curves' (Shelford, 1913) are typically bell-shaped, with their peaks representing optimal conditions for a particular physiological process and their tails representing limits of tolerance. We may define the overall range of tolerance in terms of an upper and a lower **lethal limit**; within that we may recognize a pair of inner limits, the **critical maximum and minimum** beyond which the organism, though not dead, is ecologically inviable. Within this we may recognize a narrower range of conditions, a **preferred range**; within this again an **optimum range** (Figure 6.1). Similar curves may be drawn as envelopes of the tolerance limits of a number of individuals to determine the limits to tolerance of a whole species, the original conception of Shelford (1913) who then used such species tolerance curves to explain the observed distributions of animals and plants within biogeographic regions. (We may note in passing that Shelford's tolerance curves represent a close analogue of the niche curves constructed in Chapter 4, but

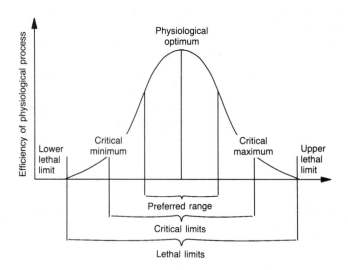

Figure 6.1 Physiological limits to tolerance. A schematic representation of the changes in efficiency of some organism across the range of some physical or chemical conditions – and the nested limits to a defined preferred range, critical limits to operation and upper and lower lethal limits.

in relation to abiotic environmental conditions rather than biological resources.)

There is in practice considerable interaction between the effects of different environmental variables, so that it is somewhat misleading to consider responses to any one abiotic factor in isolation. Tolerance to temperature for example in many terrestrial organisms is intricately bound up with tolerance to relative humidity because the physiological processes affecting temperature regulation are themselves controlled by water availability. In more general terms, one may frequently observe interdependence of tolerance to pairs or groups of environmental variables which affect the same physiological process; as a result, organisms which are stressed along one environmental gradient are less able to tolerate a wide range of conditions along other related environmental gradients as well (Figure 6.2). Once again we may draw clear parallels with niche theory – that it may be extremely misleading to consider resource relationships/resource utilization patterns along one dimension of the niche in isolation.

Such physiological limits to tolerance of a whole range of abiotic parameters clearly have a profound influence on the distribution patterns of animals and plants – and their efficiency of operation once established. Limits to tolerance, however, cannot determine which organisms do occur in a particular community; they can merely determine which species may not occur. Of all those species which have the appropriate physiological machinery to cope with prevailing environmental conditions, only one small subset is represented within any actual community matrix. To be considered, you have to get there.

6.2 ATTENDING FOR INTERVIEW: DISPERSAL

Dispersal of any organism to a given point in space depends on various characteristics of the organism itself, the relative scale of any biogeographical barriers to such movement and the relative sizes of source and sink communities. To descend to the fatuous, birds and flying insects have a higher dispersal range than slugs or springtails; plants too have seeds or fruits with markedly different dispersal range. More generally, dispersal ability is not absolute, but relates to both size and taxon. Thus any consideration of dispersal potential must be considered separately for any candidate organism; distances and dispersal times must be considered not in absolute terms, but in relation to dispersal capacity.

Within such a framework, however, the probability of any species reaching a given point in space in sufficient numbers to establish a viable population will clearly depend upon size and vigour of the source population, relative distance from source to sink – and physical size of the 'receiver' ecosystem.

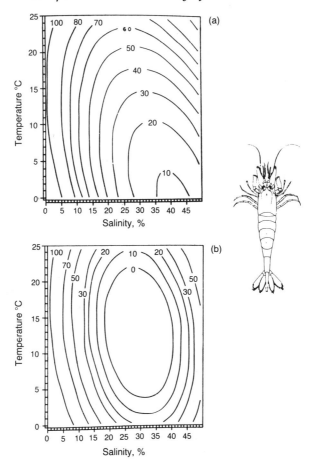

Figure 6.2 Interaction in tolerance to combinations of environmental variables. Two or more different variables may interact in determining an organism's tolerance limits to abiotic conditions such that fitness overall is reduced at the extremes of any one of the environmental factors. The Figure shows the percentage mortality of egg-bearing female sandshrimps, *Crangon septemspinosa*, at different temperatures and salinities. Figure 6.2a shows results at low concentrations of dissolved oxygen; (6.2b) shows results in aerated water. (After Haefner, 1970.)

Following MacArthur and Wilson (1967) and later authors, we may note that colonization rates of islands depend both upon island size (larger islands present a larger catchment surface/interception area to migrant individuals; Osman, 1977) and distance from the source of colonists. Extinction rates similarly vary with island size (MacArthur and Wilson, 1967) and distance from the primary source of colonists (newly established populations may be supported – or rescued from extinction

– by continued arrival of reinforcements from the source population; this rescue effect diminishes with distance from the seed source: Brown and Kodric-Brown, 1977). Although the theory of island biogeography was conceived in relation to an analysis of the dynamics of colonization and turnover of species on oceanic islands, the same basic principles may be applied to any community; after all, while it may not represent a physical island surrounded by sea, even a mainland ecosystem may be considered an 'ecological island' in that it represents a particular and discrete area of its type, surrounded by a 'sea' of different nature. It too may be defined in terms of its area, or distance from the 'mainland' of a similar system.

6.3 FINAL ELECTION TO MEMBERSHIP

Of the deluge of propagules and potential colonists continually raining down upon a community, some will be physiologically unsuited to the conditions experienced at that point in space and time (although their chance may come later; early colonists may modify the environment in such a way – through changes to its structure or chemical composition, through creation of appropriate microclimates or microhabitats – that it becomes more favourable to those who fail initially; page 95). Of those potentially able to withstand the abiotic conditions, others may be unable to establish viable populations because of the absence of food organisms (once again: at that point in space and time), because of the presence of superior competitors or predators. The community will be assembled through the chance of who gets there when and biotic interactions among those potential colonists.

But the important point to note is that the rain of propagules does not stop once a community has become established in a given area. The same rain continues throughout the life of that community; it is continually challenged by new invaders: additional individuals of species already represented in the community, individuals representing new species which have never previously been encountered by the community, and yet others: regular challengers which have merely failed to establish themselves up to that point in time.

Species within any community at a given point are continually challenged by invaders; the community itself is subject to a continuous flux of colonizations and extinctions. It is a mistake to consider any community once established as a settled matrix; it is subject to continual challenge. With such extreme potential flux there is equally resultant change in the community's composition; there is a regular turnover of membership, to offer different assortments and combinations of species. Some may prove more persistent and stable than others.

6.4 ECOLOGICAL SUCCESSION

Amongst the constant flux in species composition within communities there may be distinguished some apparent directional change of structure, an 'orderly process of community development that is reasonably directional and [apparently] predictable' (Odum, 1969). Ecological successions may be viewed as a progressive change in community composition and dynamics, whereby many of the minor changes of structure and operation accumulate over time so that the community itself develops a new emphasis, a different basis, a different 'economy'.

Clements, often viewed as the forefather of successional theory, perceived succession as a deterministic phenomenon, a development within the community proceeding directionally to some 'climax' community through a series of distinct seral stages (Clements, 1916). Succession was seen to be convergent, with communities of a wide variety of initial starting points converging on a limited number of climax states, considered characteristic of the gross geologic and climatic character of the region. Development towards such climax states was also regarded as predictable, such that if the progression was disrupted in some way and, through interference, set back a number of steps, the succeeding secondary succession would go along very much the same path again, taking the same steps and certainly converging back to the same endpoint. Although it was accepted that the process of development was a gradual one – a continuous accumulation of minor changes of composition and structure, Clements considered that it was possible to recognize within this process a series of discrete seral phases through which the community would pass, each of which might be defined almost as distinct communities in their own right.

Clements's deterministic view of community development was challenged by Cooper (1926), Gleason (1926) and others, who interpreted communities (in all cases plant assemblies; successional theory was firmly rooted in analyses of the dynamics of plant 'communities') as 'random' aggregations of individuals, existing merely as a snapshot in time at one point along some temporal continuum of change. Within such an individualistic view of community development however, with continuous replacement of a species here or a species there, these authors recognized that there must be biological processes acting to bring some order into the system – for they acknowledged that vegetational assemblies with similar histories and growing under similar environmental conditions were often very alike in composition. Presently, theory tends more towards the ideas of Gleason, based upon changes amongst aggregations of individuals rather than on higher-order dynamics at the whole-community level.

Ecological succession

The definition of apparently distinct and recognizable seral stages noted by Clements, may be derived in such models purely in relation to observed levels of interdependence between species (Whittaker, 1967). The gradual changes in environmental conditions which accompany succession could result in a number of different patterns of change in the organisms of the community, either continuous and imperceptible, or with discrete quantum shifts in community structure (transitions between distinct seral stages), depending on the ecological lability of individual species and the strength of their reliance on one another – the different levels of dependence of one part of the community upon another (Figure 6.3). Factors underlying the apparent directionality and apparent convergence of community structure to a limited number of climax states were harder to define.

Clements's views on the underlying mechanism 'driving' succession and imposing such directionality on the process were based on the idea that each community of organisms established in a given area necessarily alters the environment colonized and in so doing may permit invasion by other species – species which could not colonize the site until their arrival had been 'facilitated' by the changes wrought by earlier arrivals (Clements, 1936; Connell and Slatyer, 1977).

Any one organism, as we have established, has a limited range of environmental conditions in which it can exist, and within those, certain optimum conditions that suit it best. Yet, as any organism, or community of organisms, occupies any environment it necessarily modifies that environment – altering its physical structure, changing nutrient availability, providing new microclimates; such change may mean that the new environment suits the original colonist less well than it did, and/or that it now becomes more suitable for species which previously were unable to invade the system. Thus either the environmental conditions become so altered that the original colonist can no longer exist within the community and becomes extinct – with its place later taken over by new immigrant species, or the new colonizer, better adapted to the changed environment, actively displaces the original colonist through competition. What causes the directional change in community structure is this continued alteration of the environment. The community of organisms which occupies a particular environment at any one time has a significant impact upon that environment, which thus becomes less suitable for many of its member species and more suitable for others. This new 'community' alters the environment yet again and itself is gradually replaced, species by species.

While initially conceived as a mechanism to explain the directional change in species composition of plant assemblies – as indeed were all models of succession – the model can readily be generalized to consider succession at a more realistic community level. Occupation of a particular

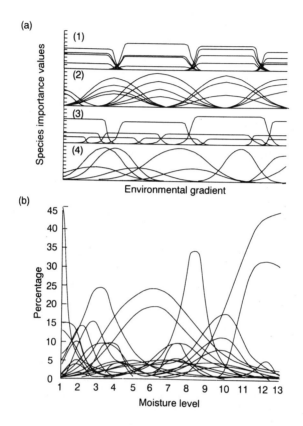

Figure 6.3 Species associations and seral structure in Succession. (a) shows four possible hypotheses of the pattern of change in abundance of different species along some environmental continuum (gradient in space or time). Panel 1: dominant species are evenly spaced and replace each other sharply at specific points; distributions of other species are strongly associated with those of dominants. Panel 2: dominant species equally spaced but replace each other more gradually; distributions of subordinate species still strongly associated with those of dominants. Panel 3: dominant species evenly spaced and replace each other sharply; patterns for all other species similar but not correlated with those of dominants. Panel 4: Both dominant and subordinate species show gradual replacement along the gradient; distributions are not strongly associated thus recognisable seral phases will not be apparent as distinct 'communities'.

(b) shows the actual distributions of individual tree populations in the Great Smoky Mountains, Tennessee, along a moisture gradient (through space, not time) (Both Figures modified from Whittaker, 1956, 1970.)

environment by some community modifies, as before, the physical and chemical environment. Such modification will of course make the physical and chemical conditions more favourable to different types of organisms – an effect primarily reflected in a change in species composition among the primary producers, whose dependence upon physico-chemical conditions is more direct. But associated with that change will be a change in the community as a whole. The species array of consumers and decomposers must also respond to the altered conditions, partly as a direct result of changes in the abiotic conditions in relation to their own tolerance limits, partly in reflection of changes in the spectrum of primary producers (and subsequently primary consumers), resulting in changes in the availability, for any specialist consumer, of particular plant or animal species as food. The process of modification continues, the spectrum of producers changes continuously – and the animal part of the community 'tracks' this change. But it is important to recognize that although the effect of succession upon the community is mediated through its more immediate effect upon the plant assembly, with changes in the animal composition of the community largely in response to changes in the composition of the producers on which they depend, the modification of the environment which leads to such a change in plant species composition in the first place is itself effected by the whole community of both animals and plants. Further, although in large part the successional change is mediated in this way through the green plants, it is not restricted to such change and other changes may accompany this larger scale movement, due to successional changes within the animal part of the community alone. Animals and plants are inextricably interrelated in any community; any process which affects or is affected by one, will influence or be influenced by the other. While succession is (regrettably, still) regarded by many as primarily a botanical phenomenon – a change in the plant array alone, whose causes and consequences may be sought in that single trophic class – such conception is as erroneous as it is misleading.

Such a model of the mechanism directing successional change fits well such processes as primary successions in the wake of retreating glaciers (Crocker and Major, 1955; Viereck, 1966) where the bare ground must be broken down and enriched with nitrogen fixed by legumes before it is possible for other species to establish, or sand-dune successions (e.g. Cowles, 1901) where the dune must be stabilized by growth of marram grass before later species may colonize. It also fits most elegantly the observed succession of blowfly species within animal carrion recorded by Fuller (1934) – a purely animal succession.

Alternative models have been proposed (see Connell and Slatyer, 1977, for a review). The 'inhibition' model interprets succession as a process where, far from facilitating the arrival of later colonists, pioneer species inhibit their invasion and establishment within the community. Once again formulated primarily to explain succession within plant assemblage, it may also be extended as a model of community-wide processes, where changes among consumer and decomposer species merely track observed changes among the producers. Since in plant competition the outcome of any interaction is partially determined by the relative sizes of the individuals concerned (resource availability relates primarily to area 'held'), small propagules of invading species will be at a competitive disadvantage to established species which already hold resources. Succession is seen as proceeding through invasion of gaps by longer-lived and more competitive species, which outlive the early colonists. Clearly such a model may apply only in later stages of succession or during secondary successions, explaining perhaps why rates of change decline in the later seral stages, and the community stabilizes at climax; it is in a sense a model of resistance. Connell and Slatyer propose one further model; this 'tolerance' model is somewhat less deterministic in conception than the others, proposing that all species present during a succession may be able to colonize the initial site. In such case, the actual sequence of colonization will be essentially stochastic (see page 106) although some species will have better dispersal ability than others and thus a higher probability of early arrival. Greater numbers of more competitive and longer-lived species will invade as the succession proceeds and as selective advantage shifts from species with high dispersal ability (r-strategists) to those with high competitive ability (K-strategists). Working within grasslands, Fenner (1978a, b, 1980) has demonstrated that species generally recognized as early colonists have markedly poorer competitive abilities than closed turf species.

The mechanisms suggested by Connell and Slatyer are clearly not mutually exclusive; indeed it is possible that all are involved during the course of any single succession: as noted, the processes defined seem to characterize different stages within the entire successional sequence. However, while it may be of interest to know the biological basis whereby one species may be displaced by another, the actual mechanism may not be fundamentally important in accounting for the directionality or convergence observed in such successions. Indeed the very properties which sparked off all such analysis of the underlying mechanisms of succession may need no biological explanation at all.

6.5 SUCCESSION AS A NECESSARY STATISTICAL ARTEFACT

The peculiar properties of directional change and repeatable convergence on a limited number of end states are not unique to succession, or indeed to biological systems; as pointed out by a number of authors (e.g. Horn, 1975, 1976; Usher, 1979), such properties are shared by a class of statistical processes known as 'regular Markov chains'. A Markov chain is a stochastic process in which transitions between various 'states' occur with characteristic probabilities that depend only on the current state and not on any previous state (Kemeny and Snell, 1960). A Markov chain is 'regular' if any state can be reached from any other state in a finite number of steps and if it is not cyclic.

The fundamental property of a regular Markov chain is that eventually it settles into a pattern in which the various states occur more or less randomly with characteristic frequencies that are independent of the initial state. It may be argued that this final 'stationary' distribution is the analogue of a climax community and that climaxes must occur by the statistical certainty that the Markov process will always settle into a stable pattern (Horn, 1975). In effect: if one may define for any pair of species a specific and, at least theoretically, measurable probability of replacement of one by the other (and by definition there must exist such a finite probability), then, from the matrix of transition probabilities we may generate for any set of species in a given biogeographic region, succession is inevitable – and a succession which will show all the properties of directionality, convergence and a fixed end-point or alternate end-states (climax, or climax cycle).

Various authors have modelled succession within different communities with Markovian processes and found a remarkably close fit (e.g. Waggoners and Stephens, 1971; Leak, 1970; Botkin, Janak and Willis, 1972; Horn, 1975, 1976, 1981; Facelli and Pickett, 1990). Table 6.1 shows the matrix constructed by Horn (1981) of the relative probability, after the demise of an individual tree of each of the dominant canopy species typical of broadleaved forest in the eastern US, that it will be replaced in time by another individual of the same species or one of the other dominant species. Based on such empirical probabilities, Horn used Markovian models to predict the future structure of a pioneer site of known canopy structure after periods of 50, 100, 150 and 200 years (Table 6.2).

Table 6.1 Transition probabilities among canopy trees of deciduous forest in New Jersey

Present occupant	Occupant 50 years hence			
	Grey birch	Blackgum	Red maple	Beech
Grey birch	0.05	0.36	0.50	0.09
Blackgum	0.01	0.57	0.25	0.17
Red maple	0.0	0.14	0.55	0.31
Beech	0.0	0.01	0.03	0.96

The probability for any given individual of each of the four main canopy species that it will be replaced in 50 years by another individual of the same species, or by a tree of another species. From Horn (1981.)

The predicted structure at steady-state (infinity) can be compared with that actually observed in an adjacent block of climax forest; the fit is indeed remarkably good.

Table 6.2 Markovian predictions of forest succession

	Age of forest (years)						Actual data from old forest
	0	50	100	150	200	α	
Grey birch	100	5	1	0	0	0	0
Blackgum	0	36	29	23	18	5	3
Red maple	0	50	39	30	24	9	4
Beech	0	9	31	47	58	86	93

Based on the empirical transition probabilities of Table 6.1, Horn (1981) used Markovian models to predict the composition of a deciduous stand at different intervals of time, starting from a pure stand of grey birch. The composition of the climax woodland (at infinity) may be compared directly with that of very old forest in an adjacent block.

From such analyses Horn (1976) concluded: 'Several properties of succession are direct statistical consequences of a species by species replacement process and have no uniquely biological basis'. The process of succession must of statistical necessity stabilize in a stationary ('climax') state; the fact that different pioneer communities might converge to the same climax could also arise merely as a statistical necessity. If a community is temporarily disturbed, something akin to the original community is restored over time; this too is a function of Markovian processes. Finally Markovian developments, like successions, are characterized by periods of rapid change followed by undetectably slow change; hence stability, in the sense of absence of change, increases tautologically as succession proceeds towards a stable state. None of these properties is necessarily of biological origin (Horn, 1976). Here then is the ultimate stochastic model – which would doubtless have delighted Gleason!

Of course this does not mean there is no biological reality about succession at all; it does however infer that many of its characteristics are not necessarily biologically determined or of special significance.

6.6 ASSEMBLY RULES

Most theories of succession or alternative models of community assembly assume some major role for biotic factors, notably competition, in determining the resulting species array. While it is accepted that factors such as the particular sequence or context of arrival of individual colonists will affect their availability for inclusion within the community matrix, the role of biotic interactions between existing community members and between them and fresh challengers is presumed the more significant. After all, given sufficient time, one might presume that any given community would have been repeatedly challenged, and at a variety of different junctures during its development, by all potential colonists. More recently, however, the relative importance of competition has been contested.

In his classic studies of the distribution of land bird species across the Bismarck archipelago near New Guinea and the species composition of 'replicate' assemblages on different islands, Diamond (1975) came to the conclusion that there were distinct patterns of association between species – both positive and negative. These relationships were not simply restricted to associations between different trophic levels (that a species cannot exist within a community if its food species are absent) but were apparent between organisms of the same trophic class. Examining the distribution patterns of a potential 147 species of land birds over some 50 islands, Diamond became convinced that there were constraints on species combinations and proposed a simple set of assembly rules. Diamond asserts:

> 'If one considers all the combinations that can be formed from a group of related species, only certain of these combinations exist in nature.
>
> Permissible combinations resist invaders that would transform them into forbidden combinations.
>
> Some pairs of species never coexist, either by themselves or as part of a larger combination. However, some pairs of species that form an unstable combination by themselves may form part of a stable larger combination; conversely some combinations that are composed entirely of stable sub-combinations are themselves unstable.'

Diamond argues that much of the explanation for these 'assembly rules' has to do with competition for resources and with the harvesting of resources by permitted combinations in such a way as to minimize the unutilized resources available to support potential invaders. He claims that 'communities are assembled through selection of colonists, adjustment of their abundances and compression of their niches so as to match the combined resource consumption curve of all the colonists to the resource production curve of the island' (Diamond, 1975). In other words potential colonists may establish themselves within a community if the resources they require are not already exploited fully by existing community members; adjustment of niche position and breadth in response to potential or expressed competition from neighbours on the resource continuum results in a set of species occupying the available resource spectrum with minimum overlap. Within their defined 'section' of the resource continuum, individual species increase in population number to the level permitted by resource availability within their expressed niche. Total community consumption matches resources available across the continuum.

Such definition: of a group of species adjusting amongst themselves individual niche position along some resource continuum, relative abundance – interacting strongly with other members of a defined unit, less strongly with community members beyond that cluster – such definition surely echoes our earlier recognition of guild structure within communities. Some of Diamond's 'units' of structure are multi-species associations and do indeed comprise true guilds as we defined them in Chapter 5 – suggesting at the same time a real functional role for such guilds in structuring the effects of competition on species change within the community.

Publication by Diamond of his 'assembly rules' provoked a storm of controversy and heralded a heated debate – a debate almost unprecedented in the apparent entrenchment and hostility of the opposing camps (Connor and Simberloff, 1979, 1984; Gilpin and Diamond, 1984; Gilpin *et al.*, 1984). Connor and Simberloff's original challenge in 1979 argued that Diamond's assembly rules were unproven in that he had failed to test his observed patterns of species occurrence against an appropriate null hypothesis. Testing the distributional data themselves against the pattern of co-occurrence predicted by random assembly, they argued that most of Diamond's observed patterns might be expected were the species concerned distributed randomly on the islands. Gilpin and Diamond responded by criticizing Connor and Simberloff's selection of null hypothesis, claiming it to be inappropriate – the battle of words swung back and forth. In essence, the debate focused on two issues, one explicit, one less so: that of the selection of an appropriate null hypothesis against which to test Diamond's recorded patterns of distribution – and the deeper issue of whether or not competition plays such a crucial role as Diamond proposed in the organization of community structure.

Both arguments continue to rage. Thus Wright and Biehl (1982) confirmed that the criteria adopted by Connor and Simberloff in their analyses (Connor and Simberloff, 1979) were too severe and will almost inevitably lead to type II errors – in rejecting patterns actually significant; Ryti and Gilpin (1987) offer a new methodology for such analyses to overcome such problems. Diamond's assembly rules have subsequently been modelled and tested on a variety of other communities (e.g. Haefner, 1981; Case, 1983), with results certainly strongly suggestive of non-random patterns of association and thus at least indicative of a probable role of interspecific competition in structuring the species composition of communities. In his analysis of the patterns of co-occurrence observed amongst lizard communities of different islands in the Sea of Cortez, Case (1983) demonstrated very convincingly that niche overlap between recorded combinations was throughout substantially lower than that which would arise in lizard assemblies drawn at random from the common species pool; mean overlaps in resource use recorded by Pianka in his analysis of guilds of desert lizards (Pianka, 1973) can likewise be shown to be consistently lower than those that would result from randomly constructed communities (Figure 6.4).

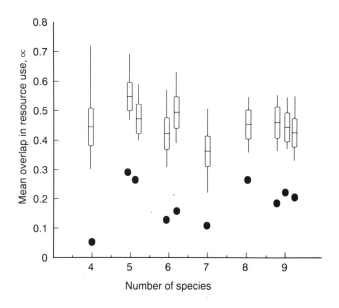

Figure 6.4 Niche relationships among guilds of desert lizards. Average resource use overlap for each of the ten lizard guilds studied by Pianka (1973) is plotted against the value of overlap to be expected in communities assembled at random from the same species pool (calculated by Lawlor, 1980). Observed values for average overlap consistently fall below those for random assembly.

But perhaps the most damning indictment of the entire original controversy, and yet at the same time the most damaging refutation of the value of analyses of distributional data, is provided by Hastings (1987). Using computer simulations Hastings modelled communities assembled under strong competitive influence (and thus known to have been ordered by a process involving strong competitive interactions) and others constructed according to simple random assortment. Hastings demonstrated that the co-occurrence patterns produced by the competitive models did not differ significantly from those developed under random assembly – even though in this case we know competition was involved. If under such controlled circumstances we cannot show significant differences in co-occurrence patterns of communities constructed under regimes of competitive or random assembly, what future is there in the battle over the interpretation of observed patterns of species association in real world communities? Yet again, in an echo of our comments on page 34, we are unable to infer anything about competition past and its possible role in community assembly processes from present-day distributional data.

We are presented yet again with a *fait accompli* – and have no knowledge of the past history. Indeed one of the biggest drawbacks with any of these analyses is the lack of documentary evidence on what has gone before. We can never know for certain whether a particular species combination is never represented in nature because some 'forbidden species' is regularly excluded – or whether that species has by random historical chance never in fact ever reached that community to mount a challenge.

Pimm and his various co-workers seem to have a remarkable talent for searching out the 'perfect' example (Crowell and Pimm, 1976; Diamond, Pimm, Gilpin and LeCroy, 1989; page 65 above); the acid test of whether or not competition does indeed play a significant role in the structuring of species assemblies must be an analysis of communities where the entire sequence and pattern of colonization is known, against which final patterns of species co-occurrence may be adjudged. Such documented history may be compiled for the exotic avifauna of the Hawaiian islands, an assemblage of non-native species introduced to the islands in the nineteenth and early twentieth century by well-intentioned 'Acclimation Societies', in an attempt to enrich the existing assemblage of bird species (or perhaps introduce some element of familiarity into the community?) (Moulton, 1985; Moulton and Pimm, 1983, 1986).

The remnants of native Hawaiian forests harbour the remains of a large and diverse endemic fauna – and relatively few exotics. Forests below about 600 m, however, are themselves largely composed of non-native species; introduced bird species are commonly found in such areas, while native species are rare or absent. Natural immigration is virtually nil because of the degree of isolation of the islands; almost all the species of the lowlands originate from deliberate human introduction – and the history of those introductions is on record. Different numbers of species have been introduced to each of the six main islands (Kauai, Oahu, Molokai, Lanai, Maui and Hawaii); amateur and professional ornithologists have undertaken regular surveys of the birds on each island over the last century so that it is possible to estimate, at least to the nearest decade, when each species was introduced and if and when it became extinct (Moulton and Pimm, 1986). These data, together with information about present-day patterns of distribution, and measurements of body size and culmen size, may be interrogated to consider (1) the overall rate of extinction, (2) the relationship between ecological similarity (as evinced by body- and mouthpart sizes) and extinction, as well as (3) relationships between extinction and taxonomic relatedness within the introduced avifauna.

In their analysis of this delightful 'natural experiment' Moulton and Pimm demonstrate that the probability of extinction of any introduced species declines in direct proportion to the morphological difference between a species and its most similar congener. Of 18 recorded introductions of congeneric pairs to a single island, there were three cases where both members of the pair became extinct, nine cases where one member of the pair became extinct and six cases where neither species has as yet disappeared from the community. If interspecific competition had been responsible for some of the extinctions observed we might expect that morphological differences (as a 'concrete' reflection of ecological differences) might be greater among the pairs of coexisting congeners still surviving than amongst those pairs from which one species had been eliminated. In only one of the pairs still coexisting on a single island was the difference in beak size less than 10%, and the mean difference was 22%; by contrast, in six of the nine pairs where one species had become extinct, difference in beak size was less than 10%, and the mean difference in this species set was only 9% (Figure 6.5).

Despite such evidence for interaction at the species scale, patterns of distribution at the community level showed no clear evidence of overdispersion. Moulton and Pimm attribute this apparent contradiction of their earlier clear evidence of competitive interaction to a dilution effect. In essence analyses within congeneric pairs are carried out within groups of morphologically similar species that share an evolutionary history; analyses at the community level span a far wider range of morphological variation over a range of less closely-related taxa. Species in the former combinations are likely to compete strongly – and Moulton and Pimm's evidence indeed suggests that they do so. Many species in the wider community context are likely to compete weakly if at all and thus might not be expected to contribute significantly to each other's risk of extinction or to discontinuities in morphological similarity. When one dilutes 18 populations of congeneric pairs into the entire species set of 107 populations, any effect of competition between the strong interactors is quickly masked (Moulton and Pimm, 1986; see also Grant and Abbot, 1980; Bowers and Brown, 1982; Diamond and Gilpin, 1982; Colwell and Winkler, 1984).

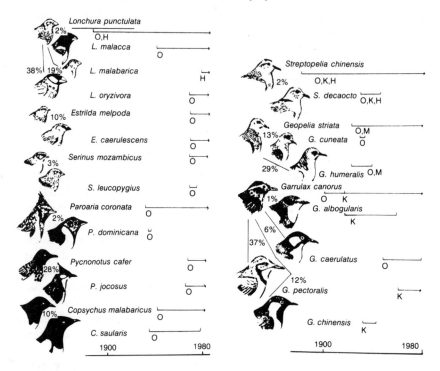

Figure 6.5 Morphological similarity between congeneric species pairs of exotic birds introduced onto the various islands of the Hawaiian archipelago and the relative success (survivorship) of pairs of different degrees of similarity. Percentages show relative difference in bill lengths of particular species pairs; vertical lines indicate approximate years of introduction and extinction. (From Moulton and Pimm, 1986.) Islands are Oahu (O), Kauai (K), Hawaii (H) and Maui (M).

6.7 COMPETITION AND THE INFLUENCE OF INVASION HISTORY ON COMMUNITY ASSEMBLY

While all this is, even in the most uncharitable view, at least strongly suggestive that competition does play some role in determining species structure and species associations within communities, it has become equally clear through all the arguments that random, or stochastic, events may also play a significant role, both in the initial development of the community matrix and in the maintenance of that structure (Nee, 1990).

For the first, it has become clear that invasion history (the sequence and timing of arrival of different colonists in a community) has a significant effect on their chances of establishment and on the resultant community structure – an effect still detectable after many generations (e.g. Drake, 1990, 1991).

Perhaps the best evidence that invasion sequence does affect the pattern of community development in this way comes from the growing number of experimental analyses of community development in laboratory microcosms (Robinson and Dickerson, 1984, 1987; Robinson and Edgemon, 1988; Drake, 1991). Thus the effects of invasion sequence on resultant community structure were evaluated by Robinson and Dickerson (1984, 1987) using a series of water-filled beakers, inoculated with replicated subsets of species from a species pool consisting of green algae, blue-green and golden algae, ciliates, euglenoids, a rotifer and a flagellate. These species were added to the beakers in four distinct and predetermined sequences and at two different rates over a period of weeks. Robinson and Dickerson detected significant differences in community structure and species richness which could be attributed to both sequence of invasion and timing of introduction. Essentially similar results were found by Robinson and Edgemon (1988) and Drake (1990, 1991) – all of which argues cogently for an effect of invasion history on the composition of the resultant community.

However, there is also some evidence from these studies that the communities once assembled subsequently adjust to become less vulnerable to further invasion and some suggestion at least in certain cases of some 'deterministic' convergence in the end-communities. Robinson and Dickerson note that 'priority effects were found to be important in some *but not in all instances* [my italics]' and the assembled communities of their particular experimental protocol could be divided into two primary types: those dominated by *Ochromonas* and those dominated by *Paramecium bursaria*. Significantly, they also note (Robinson and Dickerson, 1984) that *Ochromonas* communities were invulnerable to invasion when, after 23 weeks, the assembled communities were challenged by introduction of a set of three previously unencountered species. By contrast no clear pattern of invulnerability was apparent for *Paramecium*-dominated communities, which altered in composition following establishment of these later invaders. Such observations suggest there may be some pattern of convergence, despite the influence of invasion history (two clear end-communities, despite the wide range of experimental treatments) and that indeed there may be a unique end-community invulnerable to invasion and resistant to further change. The evidence for such a claim is, however, tenuous and such convergence is not reported by other authors (e.g. Drake, 1990, 1991), who emphasise differences in relation to invasion history.

As we have had occasion to note before, conclusions from simple laboratory manipulations are not always matched in the field. In this case we may examine the species composition of a series of small seasonal pools on a freshwater marsh in East Lothian, investigated by Jefferies (1989). The delight of Jefferies's analysis resides in the fact that while these pools were subject to natural colonization from the surrounding marsh and the adjoining permanent water of a small loch, the pools themselves were identical. Formed after the war by the removal of a row of concrete pillars erected as anti-tank defences, the pools were of the same age, same shape (square), same area, same depth and had the same substrate (Jefferies, 1989). And there were fifty of them, extending in a row from the edge of the permanent loch to over 400 m distance.

Because of differences in distance from a source of colonists, and small but consistent variations in physico-chemical properties of the water, the pools sampled were organised into eight clusters in relation to distance from the loch, extent of summer drying, and links to other pools or the main loch when flooded; within each set thus derived, there still remain from four to seven pools. Faunal assemblies within each 'physical' set of pools thus derived were remarkably homogeneous with regard to distribution of 99 defined taxonomic groups. Despite some overlap, distinct and coherent communities did appear to exist.

A number of taxa were recorded on only one or two occasions in only one or two pools; Jefferies divided the 79 taxa that were more regularly encountered into those which were sampled in more than 50% of the pools in any class and those sampled from fewer than 50% (repeating the analysis separately for each 'class' of pools defined), suggesting that those taxa recorded in more than 50% of the pools of any class might be 'expected' to have colonized all such pools, while those sampled in fewer than 50% of the pools were 'unlikely' members of such communities. Overall the mean frequency of occurrence of any taxon in pools in which it might be 'expected' was determined as 79.6%; we may consider this an index of predictability or determinism of the community structure of any pool-type. The mean frequency with which any given taxon was recorded from 'unlikely' pools (included in communities of which it was not commonly a member) may be taken as some measure of deviation of community composition from what would result from purely deterministic processes, or a measure of the element of chance (Talling, 1951); for Jefferies' tank-trap pools this element of chance was measured as 10.9%.

7
A question of equilibrium

The influence of stochastic events on community structure is not apparent merely in an effect of history during community assembly. Unpredictable variation in environmental conditions may also help explain the occasional observation of persistence within communities of what Diamond (1975) would brand forbidden combinations: the coexistence of strong competitors. We considered briefly in section 2.5 the implications of extending purely deterministic models of competitive interaction to include an element of lottery competition (Sale, 1977, 1979; Warner and Chesson, 1985; Silvertown and Law, 1987) or sporadic episodes of high density-independent mortality (Huston, 1979). We also showed that coexistence of strong competitors might be facilitated by periodic fluctuations in conditions, resulting in regular reversal of competitive advantage or simply, through continuous disruption of the community's dynamics, preventing such interaction from ever running its full course to extinction. Indeed it can be shown that intermediate levels of disturbance will promote maximum levels of diversity within communities (Horn, 1975; Connell, 1978; and see page 122–3 here); provided the perturbations are neither so severe or so frequent that they actually destroy the community, the continual disruption of predatory or competitive interactions permits the coexistence within the community of powerful antagonists.

Such considerations of the potential importance of stochastic events – and periodic environmental perturbation – on community assembly and the maintenance of structure must beg the question of how common are such fluctuations in conditions. Indeed, do communities ever each equilibrium structure and dynamics – the apparent end point of their 'evolution' and development, or is that gradual process of continuous adjustment towards a final steady state always interrupted before such equilibrium is reached? If a community is continually disturbed and interrupted in this way, then many of the structural properties we might expect to find, derived from our considerations of the eventual consequences of biotic interactions in the limit, may never be observed because community processes in practice never approach that limit – and community structure and dynamics may be more characterized by the effects of stochastic events and periodic disruptions than by the end point of the gradual processes of biotic interaction and adjustment which are never given the chance to run to their logical conclusion.

Much of the ecological theory of the 1960s and 1970s was based on the supposition that present-day communities in general are in some kind of equilibrium (and that so were those studied). However, if regular challenge to such systems is more commonplace than might have been assumed then either (1) some of the communities studied, from which we have drawn general principles of structure and dynamics, may not themselves have been at equilibrium (in which case the characteristics recorded may not be a reflection of the properties of equilibrium systems, but rather reflect merely some passing temporary phase); or (2) if a good proportion of present-day systems never reach equilibrium anyway, then even if it were possible to elicit certain 'rules' of organization of stable systems, such rules or constraints might not themselves be very widely applicable.

A good overview of the issues, arguments and counterarguments is provided by Wiens (1977), Grossman (1982) and Miller (1982), with further reviews by Wiens et al. (1986) and Chesson and Case (1986). As with many issues in ecology, support may be found for both viewpoints. Wiens (1977, 1986) argues persuasively that the possibility that ecological systems may not be at equilibrium has been widely overlooked, yet that perhaps this is rather the rule than the exception. In their studies of shrub-steppe bird assemblies in North America, Wiens and Rotenberry (1980) could demonstrate no clear evidence of resource limitation (a prerequisite for invoking competition and thus biotic interaction as a structuring force); nor could they demonstrate any predictable changes either in the species composition of the entire assemblage, or the size of population of individual species within the guild as availability of resources changed. Their analyses of the changes in species composition and relative abundance suggested that the system was more powerfully affected by stochastic variation in abiotic factors – affecting each species population in isolation – than by biotic interactions between guild members.

Other evidence that many communities are not competition-structured, as might be expected for systems at equilibrium, is presented by Strong, Lawton and Southwood (1984) in their analysis of the dynamics of insect–plant systems. Although such phytophagous systems may seem something of a special case, we should note that they account for about one-quarter of all living species; from their review of the published data from a wide variety of such phytophagous insect communities. Strong *et al.* conclude that such systems are rarely structured through competitive interactions. Indeed in only 17 of the 41 published studies that they considered was there definite evidence of competitive interaction, and in the majority of instances even this involved only a small proportion of the community's members. Yet against such argument may be ranged many examples claiming to demonstrate clear evidence for the importance of biotic interactions in determining structure, transcending any stochastic disruption of the community's development; we have already reviewed some of the evidence that biotic interactions truly do exert some impact in controlling the dynamics of real communities in previous chapters.

In effect the whole argument is perhaps oversimplistic and extreme; while we should adjust our mental focus to acknowledge that many communities may not be at equilibrium, this does not imply that *all* communities are so ordered. The fact that we may establish that not all communities are at equilibrium does not lead necessarily to the conclusion that all must be non-equilibrium systems; that surely is a logical fallacy. Some communities may indeed show few effects of biotic interaction and be primarily ordered by stochastic processes, others may truly have reached an equilibrium; yet others again may show properties of both (as Jefferies' elegant attempt to quantify the relative roles of chance and determinism in regulating the species composition in the tank-trap ponds of our previous chapter clearly testifies).

As Grossman (1982) notes: 'The classification of whole systems as either deterministic or stochastic is somewhat of an oversimplification. It is clear that both types of properties may influence the abundance of species in either type of assemblage.' What is equally clear, however, as Grossman continues, is that in many instances, 'a majority of inhabitants appear to be regulated by one or other mechanism. The differentiation of these two types of assemblage is useful because the processes responsible for [for example] the coexistence of taxa are very different in these systems.' Grossman goes on to consider equilibrium models of coexistence, and non-equilibrium models such as those invoking lotteries or episodic mortality from density-independent factors, as we have done on pages 32–3 and 109. We may, however, broaden the perspective.

In equilibrium systems we might expect to find structure primarily ordered by biotic interactions (particularly competition) and coevolution between competitors, mutualists, predators and prey. We would expect to find non-equilibrium systems invasion-structured, influenced in their structure and composition far less strongly by biotic interactions (with, amongst those, predator–prey relationships the more apparent), organized instead primarily by independent interactions of species and species populations with abiotic factors. Such systems may be understood purely by looking at the separate interactions of each individual population with abiotic limiting factors, ignoring biotic interactions with other species since these will either be non-existent or of minor significance in relation to the overriding dominance of abiotic factors. How then may we distinguish the two types of system? Are there any predictable situations which would lead inevitably to equilibrium or non-equilibrium dynamics? Are there any obvious characteristics by which we might recognize, from their expressed structure rather than from any detailed knowledge of their underlying process, predominantly invasion-structured or biotically-ordered communities?

The implications for the final structure of communities primarily constructed through simple invasions and those where each addition to the community matrix is followed by a period of co-evolutionary adjustment before exposure to the next challenge, have been explored by Rummel and Roughgarden (1983) and Roughgarden (1986); sadly however, the structure finally expressed reveals little trace of the specific assembly process: organization of species in resource space is substantially the same for model communities constructed by either process. There seem to be no short cuts.

All we may suggest perhaps is that, as entire assemblages or as components within a community of mixed assembly, invertebrates and annual plants might be more strongly influenced by stochastic processes. Such organisms are commonly extreme r-strategists with little or no density-dependent feedback on population numbers; thus populations show little temporal or spatial stability. Generation times too are relatively short; with many generations completed within the period of any gross perturbation, populations will experience immediate selection pressure from instantaneous conditions and will not have the opportunity to respond in any evolutionary sense to 'cope' with such perturbation (above, page 32). By contrast, more extreme K-strategists, (long-lived plants, vertebrates) have greater likelihood of reaching equilibrium; populations are in general more stable and thus more likely to be involved in sustained interaction and longer generation times, more likely to span the periodicity of minor perturbation permit evolutionary adaptation to such regular disturbance.

7.1 DETERMINISM, STOCHASTICISM AND CHAOS

Recognition that random, stochastic events may play a major role in structuring the dynamics of communities subject to frequent disturbance, and thus perpetually in disequilibrium, challenges to a degree the power of deterministic models of community dynamics to explain the observed patterns of structure and dynamics of any ecological community or to predict future change with any measure of confidence. But even if that

Determinism, stochasticism and chaos

were not the case, even in equilibrium systems deterministic models may in certain circumstances have limited predictive power.

Meteorologists abandoned some time ago their practice of offering long-range weather forecasts, recognizing that the accuracy of their predictions for more than a very few days ahead was little better than that of traditional folklore (that rooks build their nests lower in the trees in years when one may expect a stormy spring and summer or that a fine fruit and seed crop on shrubs and trees augurs a hard winter ahead). It was not the validity of their models that was to blame; years of accumulated data have honed very sophisticated and precise models for predicting such weather systems. Rather it was the recognition that tiny imprecisions of measurement of some of the many variables fed into such models, or the accumulation of unmeasurably small, natural changes in their value over time, could combine over many generations of the model such that its accuracy very suddenly changed to offer predictions indistinguishable from random sequences (usually after some two or three generations, predicting two or three days ahead).

More generally, deterministic models of a number of physical and biological phenomena may fail to predict accurately the dynamics of the process modelled beyond more than a few generations. Tiny changes in some of the factors affecting some dependent variable may accumulate in their effect over time so that the model can no longer project the true dynamics of that variable; commonly, the accuracy of prediction remains extremely good for a (finite) number of generations and then suddenly, and catastrophically, fails.

Yet this is not necessarily a failure of the model itself; nor does it suggest that the natural process being modelled is unpredictable or random. Chaotic dynamics in any natural system may be entirely deterministic. Schaffer and Kot (1986), offer a useful summary of the characteristics of such chaotic systems (see also May, 1986b):

1. The behaviour of the system studied (or of the equations used to model it) is entirely deterministic; there are no random inputs.
2. Chaotic systems exhibit sustained motion. They do not settle to simple equilibria or limit-cycles, but oscillate between values which never repeat and are often highly irregular.
3. Chaotic systems characteristically have an extreme sensitivity to initial conditions. The behaviour of both the natural system and any deterministic model mimicking its behaviour are such that were one to follow the system through any number of repeated 'runs' from the same starting point, the developing trajectories of each projection would diverge exponentially from each other. Tiny differences in the initial conditions, so small as to be undetectable, are amplified; since one can never define a system's current state with

infinite precision, long-term prediction of its future behaviour from its current state becomes impossible.

It is this intense sensitivity to starting conditions which is so critical and results in the expression of chaotic dynamics; and by no means all physical or biological processes are so sensitive. Many deterministic processes – indeed probably the majority – are, by contrast, extraordinarily robust with entirely predictable outcomes; but there is a dawning awareness amongst ecologists that there is a small, but important, subset of processes or systems which are extremely sensitive to initial conditions in this way and whose dynamics are not so easily predicted. We do not as yet know how widespread may be this extreme sensitivity to initial conditions in the real world (but see Berryman and Millstein, 1989). Yet the implications are considerable: for when we observe in the real world that some parameter of an ecosystem varies in an irregular and unpredictable way (it might be species composition of a community or size of some component population of that community) such behaviour does not necessarily imply that that same parameter is in fact controlled purely by stochastic variation.

Such processes, despite the chaotic nature of their dynamics, are far from random; the problem then arises of distinguishing such deterministic chaos from genuinely random behaviour. When one observes apparently chaotic behaviour in population or community dynamics how may one determine whether such dynamics are entirely stochastic or are the outcome of genuinely deterministic processes? The first steps towards establishing a methodology for such resolution have been taken by Takens (1981) and Sugihara and May (1990), and some evidence for genuinely chaotic dynamics has since been detected in the population cycles of Canadian lynx and of microtine rodents (Schaffer, 1984, 1987), the periodic outbreaks of pests such as *Thrips imaginis* (Schaffer and Kot, 1985a) and childhood diseases such as measles (Schaffer and Kot, 1985b; Sugihara and May, 1990; Sugihara, Grenfell and May, 1990).

8
Species Diversity

One important dimension of the ecological community to which we have not yet paid explicit attention is that of **species diversity**. The diversity of any community is in large part a function of the total number of species it may contain (**species richness**), but also of the distribution of individuals between those member species (**equitability**): a community consisting of 19 different species all present in low numbers and a twentieth species which is extremely abundant, dominating the entire structure and dynamics of the resulting assembly, may be considered less diverse than an equivalent system in which the same 20 species are all more equally represented within the community. In this chapter we will attempt to address the question of what factors influence diversity and determine both the number of species present and their relative abundance.

(It is of course virtually impossible to account for all the species within any community and to be certain that one has accurately determined the total number of species represented or their true abundance. In practice therefore, estimates of the diversity of any community are based upon a restricted sample of the community's members; estimates of species richness and equitability are recorded in relation to the number of individuals sampled and are thus a function of the number of species present and their relative abundance *within a defined sample of individuals* (see for example Pielou, 1969; Hill, 1973; Magurran, 1988).)

It is apparent that any general theory of diversity must embrace two kinds of constructs: what Brown (1981) has termed 'capacity rules' and 'allocation rules'. Capacity rules define those characteristics of any given environment which affect its capacity to support life: influencing broadly the total number of organisms that may be sustained, irrespective of their kind; allocation rules determine the inherent 'divisibility' of those resources and the ways in which the available resources are subsequently partitioned amongst species. In this sense, allocation rules are clearly a function of those various biotic interactions – competition, predation, etc. – which influence the way resources are partitioned; in effect the 'rules' of resource partitioning already considered in Chapters 4, 5 and 6. Their implications for determining observed patterns of relative abundance within real communities are considered further on page 126. What determines environmental capacity is more difficult to define; (nor in practice do we actually have any real evidence that natural communities are necessarily filled to saturation. The considerations of the last chapter indeed suggest that regular disruption of community dynamics may prevent many communities from reaching such equilibrium.)

8.1 FACTORS AFFECTING SPECIES RICHNESS

Let us start with a few simple observations: it is noticeable, comparing otherwise equivalent communities, that there appears to be a gradual increase in species diversity from the Arctic to the equator. Further, within any one geographical zone, we may record an increase in species richness through ecological succession. Richness of species increases as a function of ecosystem size (the species area relationship of MacArthur and Wilson, 1967; for fuller discussion see Diamond and Mayr, 1976). Finally, species richness appears to increase, even within established communities, over time.

Over the years, a number of hypotheses have been put forward to try and account for such observations and offer a general theory of diversity. The most general of such models merely proposed that diversity is a simple function of time: all communities tend to diversify with time, therefore older communities will be more species rich than young ones. In effect this simply reiterates an observation; it offers no explanation of what permits the observed increase in diversity or what may ultimately limit its expansion. It is also a statement of the self-evident. Although equitability may alter within established communities through adjustment of the relative abundances of the member species, species richness itself can only increase through immigration of new species to an area, specialization of existing species, or evolution; all necessarily take time. Such reaffirmation of a general observation takes us no closer to understanding what sets the upper limit to the diversity that may be expressed within any system, what factors may determine its ultimate capacity. We must seek instead theories which offer some candidate for determining environmental capacity and the potential divisibility of those resources (that element of divisibility inherent to the resources themselves rather than the final consequence of population interaction).

Two main contenders have been proposed for environmental features that might indeed determine capacity: productivity of the system, and spatial complexity or heterogeneity. Brown himself (1973, 1981) has assumed throughout that environmental capacity must be determined by productivity – and there is no doubt that differences in diversity between similar communities, or within a single community over time, are commonly associated with a difference in the level of primary production (e.g. Cody, 1974; Brown and Davidson, 1977; review by Brown, 1981). Equally, the relationship is not always consistent; other examples may be cited where low-productivity sites have higher species richness than sites of higher productivity. Low-productivity sites may limit the abundances of all populations to levels at which they do not strongly interact. With higher availability of resources, certain species may greatly increase in abundance to monopolize those resources, excluding others or so dominating the community's composition that diversity falls. A classic illustration of this is provided by Yount (1956) in his analyses of species number and system productivity at Silver Springs in Florida; his results show clearly that in this ecosystem low-productivity sites were associated with higher species richness.

This difference in the response of species richness to increased production may depend on the form of the increase in productivity. MacArthur (1972) suggested that an increase in the abundance of all resources brought about by an overall increase in production might indeed lead to an increase in diversity or species richness; by contrast, increase in production within one small part only of the total resource spectrum might lead to a decrease in diversity because the community would then become dominated by those species superior at exploiting that particular range of resources. Illustration of the complexity of the relationship and the fact that an increase in productivity may, depending on its form, be accompanied by either an increase or a decrease in species richness may be drawn from the work of Abramsky (1978) on small mammal assemblies. Abramsky manipulated the abundance of food in 1 ha plots of shortgrass prairie in Central Plains, Colorado, with the addition to experimental plots of alfalfa pellets and whole oats; in a second experimental treatment, productivity of other plots was enhanced by the application of water and nitrogen fertilizer. In the first treatment, the small mammal species naturally inhabiting the manipulated food plots did not themselves respond in any way to the supplementation of resources, but a new species, *Dipodomys ordii*, a specialist seed-forager, invaded the plot and increased the diversity expressed. In the fertilized plots, the increase in production was associated with vegetational growth and thus major change in the structural characteristics of the environment; in this case two new species colonized the experimental plots, but other small rodent species resident in the area largely avoided the treatment and in consequence diversity declined (Abramsky, 1978).

If the diversity of organisms within a community is restricted in some way by limits in the inherent divisibility of available resources, species

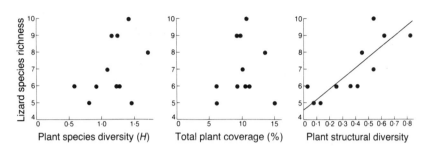

Figure 8.1 The observed relationship between the species richness of desert lizard assemblages in the south-west United States and three measures of vegetational structure: diversity, percentage ground cover and structural diversity. (Data from Pianka, 1973.)

richness might respond to an increase in structural heterogeneity which effectively 'creates' a wider potential diversity of available niches. The more complex an environment the more finely its resources may be divided and thus the richer may be its fauna and flora. Correlation between structural heterogeneity and species richness has been demonstrated in a number of studies (although sadly, as so often, studies restricted in analysis to a single taxonomic assemblage). MacArthur and MacArthur (1961), amongst others, have described for example a close relationship between bird species diversity of woodlands in North America and vegetational structural complexity. In the same way, numbers of Heteroptera and Coleoptera recorded in different communities have also been shown to be closely related to architectural complexity of the vegetation (Southwood, Brown and Reader, 1979), and the species richness within guilds of desert lizards in the south-west United States has been shown to be more closely correlated with structural diversity of the desert vegetation than with either total plant cover or plant species diversity (Pianka, 1967; Figure 8.1).

The problem once more appears to be that such relationships are not always so clearcut or consistent. Further, the argument is somewhat circular; an increase in species richness may itself produce an increase in structural complexity, as animals and plants provide habitats for others or produce diversity in micro-environmental pattern.

Neither environmental productivity nor structural complexity seem sufficient on their own to account for observed patterns of species richness; but this is not to discount them. Both may well be amongst a more complex set of contributing factors in establishing environmental capacity and there is, as we have seen, some evidence for each. What we have not yet established, however, is that the diversity expressed in any given community relates to that environmental capacity in any simple way in the first place; are communities filled to capacity, or may the development of diversity be arrested before communities become saturated?

8.2 ARE COMMUNITIES FILLED TO CAPACITY? SPECIES RICHNESS BELOW AND BEYOND SATURATION

Perhaps our search for explanations of patterns of species richness based upon the factors limiting environmental capacity is doomed to failure; perhaps our failure to demonstrate any clear correlations between species richness and productivity or spatial heterogeneity merely reflect the fact that communities are not filled in relation to their capacity. And if communities are not filled to saturation, a search for control of species number or relative abundance by biotic interactions such as competition will also be doomed; for in unsaturated communities the full intensity of such interactions will not be expressed.

Indeed we invoked those very arguments on pages 81–2 in exploration of the lack of apparent constancy in resource partitioning among bracken-feeding insects (Lawton, 1982, 1984) – suggesting that the lack of regularity of pattern in resource relationships described for bracken communities in England and in North America might in some part be due to the far lower species richness of the North American communities. If the community is not filled to capacity, biotic interactions amongst its members may not have been sufficiently intense to enforce the same pattern of resource division apparent in the richer UK bracken community. Lower species number does not, however, necessarily imply that such communities are unsaturated. What does suggest that the bracken community in North America is not filled to capacity is the recognition of vacant niches. Lawton defined for bracken-feeders a number of potential niches in terms of the way they make use of the plant, and where on the plant they might be found. (Thus the different insects might be chewers, suckers or miners, or might form galls within the plant; they might exploit the pinnae, rachis, costae or costules.) If a two-way table is created to define the potential niches available, and occupation of these various niches scored for both English and North American communities, it becomes very clear that many niches in the New World communities are left unoccupied; that these niches are not simply untenable is clear from the fact that many of these same niches are occupied by members of English bracken communities. We are forced to the conclusion that there are genuine vacancies and that the North American communities are not fully saturated. Reviewing the evidence available from other studies of phytophagous insect communities, Lawton (1982, 1984) and Strong, Lawton and Southwood (1984) suggest that such systems in general appear very rarely to be saturated.

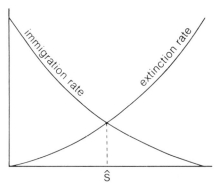

Number of species on island

Figure 8.2 MacArthur and Wilson's models for the equilibrium number of species on islands. The equilibrium number of species $\hat{S}$ is determined by the intersection of curves representing the rates of colonization and extinction. Rate of establishment of new species falls as the number of species already present on the island increases; rates of extinction rise. (After MacArthur and Wilson, 1963, 1967.)

In such case, what does determine the species richness observed? Is it merely that we are observing the community at one point in time during a progressive, slow increase in diversity – and that we are observing it before it has reached a final equilibrium – or may species richness indeed have reached a limit below that at which it would have been arrested by the limits of capacity?

The very development of a community, as we have established in Chapter 6, is a function of continual colonization and extinction – generally with the rate of colonization exceeding that of extinction. If for purely mechanistic reasons the rates of colonization and extinction came into balance during this development of the community, then species richness would be fixed, even though this might be at a level well below that at which the community would theoretically be saturated. Thus, if for some reason rate of extinction in a particular community was very high, or rate of colonization peculiarly low, the community might become balanced with a relatively lower number of species, irrespective of its theoretical capacity.

In their analysis of the way in which rates of colonization and extinction might influence the species richness on islands, MacArthur and Wilson (1967) showed that colonization and extinction rates were directly related to the number of species already present on the island. The rate of immigration of new species to any island decreases as the number of species on the island rises (Figure 8.2): as more and more of the potential

Are communities filled to capacity?

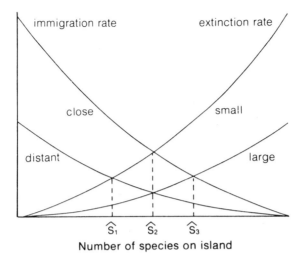

Figure 8.3 Variations in equilibrium species number for a given island, dependent on size and distance from a source of colonists. The relative numbers of species on small, distant islands ($\hat{S}_1$) and large close islands ($\hat{S}_3$), predicted by the MacArthur and Wilson Equilibrium model. The number of species on small, close, or large, distant islands is intermediate.

mainland colonists are found on the island, fewer of the new individual colonists arriving will constitute new species; most will belong to species already represented on the island. The rate of extinction likewise increases inevitably with an increase in the number of species on the island. That particular species number at which colonization and extinction rates are equal determines the equilibrium species number for the island. New species may still arrive, existing species may continue to become extinct and thus species composition may continue to change; but the actual number of species present at any time is inevitably fixed by that equilibrium between rates of colonization and loss.

MacArthur and Wilson also noted however that rates of colonization and extinction might vary with both island size and distance from a source of colonists (Chapter 6). Thus depending on the particular values of these parameters, rates of colonization and extinction might come into balance at different equilibrium species number (Figure 8.3). Such considerations may also affect the equilibrium species number expressed in any community. As we have already noted, although it may not represent a physical island surrounded by sea, even a mainland ecosystem may be considered an ecological island in that it constitutes a particular and discrete area of its type, surrounded by a 'sea' of systems of different type, often just as inhospitable and hostile to members of the focal community as a real ocean. For such mainland communities, too, rates of colonization and extinction may vary with 'island' size or distance from a source of potential colonists; diversity in any community may be determined by an equilibrium between rates of colonization and extinction – and different communities may express a different equilibrium species number even though their environmental capacities might be identical.

MacArthur and Wilson's initial theory is grossly oversimplified; as noted by Simberloff (1978a) and Gilbert (1980) it is essentially stochastic and takes no account of species interactions (see also the review by Williamson, 1981). In addition MacArthur and Wilson made the further assumptions that immigration rates, while changing with distance from a source of colonists, were independent of island size, and that extinction rates were a function of island size alone and not affected by isolation. As we have already noted, larger islands are likely to 'intercept' a higher proportion of potential colonists (Osman, 1977), and populations in danger of extinction may be more easily 'rescued' on islands closer to a source of colonists by arrival of additional individuals of the same species (Brown and Kodric-Brown, 1977). But such modifications merely affect the shape of the colonization and extinction curves and do not affect significantly the basic conclusions (beyond the fact that relative turnover rates on large and small islands are reversed; a reversal which resolves the paradoxical predictions of the original model of MacArthur and Wilson that turnover rates should be higher on larger islands).

While balanced rates of colonization and extinction might, due to 'island' size or isolation, result in an equilibrium species number below that predicted by environmental capacity, levels of diversity higher than those strictly sustainable may be maintained in any community by periodic disturbance, which prevents biotic interactions between potentially exclusive pairs of species ever running their full course (above, page 109). Just so long as the severity or frequency of perturbation is not so high as to cause population extinctions through the imposition of major density-independent mortality, periodic disturbance may prevent the community from ever reaching equilibrium and may permit the continued coexistence of strong antagonists (Connell, 1978; Miller, 1982; Warner and Chesson, 1985; Silvertown and Law, 1987).

Non-equilibrium models of community structure were considered in some detail in the last chapter and we need not rehearse these arguments again in detail. A classic illustration of this enhancing role of intermediate levels of disturbance on community diversity is, however, provided by Sousa (1979) in a study of algal communities in the intertidal zone of rocky shores in Southern California. Sousa assessed the diversity (species richness) of encrusting algae on boulders of different size, arguing that wave action disturbs small boulders more often than it does large ones. This initial (and rather crucial!) assumption was checked by estimating from photographs the actual probability that a boulder of a given size would be moved during the course of a month; on this basis Sousa defined three distinct size-classes of boulder with markedly different probabilities of movement. A class of small boulders (requiring a force of less than 49 N to move them) had a probability of 42% of being moved due to wave action in any month; an intermediate class (requiring a force of 50–294 N to displace them) had a mean probability of movement of only 9%. Finally, the class of larger boulders (over 294 N for displacement) were relatively immovable, with a monthly probability of movement of less than 0.1%. Measures of species richness did indeed show marked variation in relation to the size of boulders and thus their susceptibility to disturbance. In every sampling period, the smallest boulders had the lowest diversity of species associated with them – disturbed so regularly that community development could never reach saturation. The middle sized boulders, exposed to intermediate levels of disturbance, had however a significantly greater diversity of algal species associated with them than did the large, essentially static rocks (Sousa 1979; Table 8.1).

Table 8.1 The effects of intermediate rates of disturbance in enhancing diversity within communities

Census date	Size of boulder (as force, in newtons, required to displace it)	Species Richness Mean	Range
October/November	< 49 N	1.8	1–4
	50–294 N	3.6	2–7
	> 294 N	2.4	1–6
May	< 49 N	1.7	1–5
	50–294 N	4.0	2–6
	> 294 N	3.3	1–6

The table shows the number of species of algal communities associated with intertidal boulders of different size and thus different frequency of disturbance. Mean species richness is lowest on small boulders subjected to regular disturbance, and higher on established boulders which are rarely if ever displaced. Diversity is highest however on boulders of intermediate size, subject to intermediate levels of disturbance. Winter and summer census values are combined for 1975, 1976. (From Sousa, 1979.)

8.3 CONTROL OF SPECIES RICHNESS: A UNIFYING THEORY

Sanders (1968) suggested that both in a latitudinal progression from the arctic to the tropics and in any successional sequence, purely physico-chemical, abiotic parameters of the environment become less and less important in determining community structure, while the role of biotic interactions increases dramatically (in effect recognizing that the relative influence of stochastic density-independent factors and biotic interactions between organisms altered along such gradients). Sanders claimed that this increase in the importance of biotic interactions within a community's dynamics would facilitate the development of a higher level of diversity. Sanders, however, could not explain why certain communities might be more structured by physical or biological interaction.

We have already noted, however, in Chapter 7 that there may indeed be communities whose dynamics are more ordered by population interactions while others may be more powerfully influenced by abiotic relationships; these last are characteristically those where stochastic variation in environmental conditions is pronounced, so that conditions may vary widely – and unpredictably. Such variations in conditions may in their own right be so significant that their effect upon the population dynamics of individual species eclipses the continued influence of any competitive or predatory interactions with other community members; alternatively, frequency of disturbance may be such as to create non-equilibrium dynamics within which biotic interactions may contribute nothing at all to the community's dynamics. Environmental severity, variability and predictability may thus determine for any system whether it may show dynamics dominated by biotic interaction or by stochastic variations in environmental conditions. With such meticulous logic we have rediscovered the wheel.

Slobodkin and Sanders (1969) argued, as we have here, that the species richness of any community is a function of the severity, variability and predictability of the environment in which it develops. They claimed that any environment may be simply characterized in relation to these three major principles, so that, for example, environments might be classed as favourable, constant and therefore predictable; favourable, variable but predictable; severe, variable and therefore unpredictable; etc. Figure 8.4 expands these potential combinations and offers examples of the different classifications. As stochastic variation in environmental conditions decreases with constancy or predictability, or severity of the environment declines, so diversity may increase; the level of biotic interaction also increases, but more as a consequence of increased diversity than a cause.

No organism may colonize an environment – or have any prospect of maintaining a stable and persistent population – if that environment is severe, variable and unpredictable in space or time. As severity decreases or predictability increases, so the prospects for successful colonization improve. Thus even a severe environment may be colonized, by physiological adaptation, as long as it is constant, or at least predictable; likewise the population dynamics of organisms in such environments are less likely to be disrupted by periodic disturbance if the environment, however severe, is constant, or if variation is predictable. Ultimately in

By combination environments may be				With, as example
F	C	and ∴	P	Tropical rain forest, coral reef
F	V	but	P	Temperate climax woodland with regular seasonal fluctuations
F	V	but	UP	
S	C	and ∴	P	Hot springs
S	V	but	P	Arctic tundra
S	V	and	UP	Chemically-polluted stream Boulder beach

Figure 8.4 Slobodkin and Sanders (1969) classify environments on the basis of their severity, constancy and predictability.

favourable and predictable environments, conditions are 'easier' and within the tolerance ranges of a whole host of potential colonists; the influence of extreme conditions or extreme variation in conditions is lost completely and thus the community's dynamics are no longer dominated by stochastic changes in such conditions. By the same token, perturbations are reduced, thus the community may develop towards equilibrium; biotic interactions become the most important influences on the dynamics of its constituent populations.

Thus diversity is seen to increase as environments become more favourable or more predictable. Movement from the tundra to the tropics, or from a pioneer to a climax sere, will indeed be reflected in a decrease in severity and unpredictability and will be accompanied by the observed increase in species richness. A decrease in environmental severity, an increase in constancy or predictability will themselves also be matched in all probability by parallel increases in productivity and structural complexity; thus observed correlations of species richness with production or spatial heterogeneity may simply be first order reflections of this more fundamental pattern.

Craik (1989) has argued that there may be some circularity in the original definition of a severe environment – as one which is seen to be colonized by relatively few organisms. He notes that at a time in evolution when multicellular life was confined to the sea, freshwater was in effect a completely abiotic environment; at an earlier stage still the developing oxygenation of the atmosphere must have presented a severe challenge at a time when life was anaerobic. He argues that extreme conditions *per se* have not been enough to pose a barrier to occupation by a diversity of living organisms: that over evolutionary time organisms have the opportunity to adapt to any extreme of environment. Where

Slobodkin and Sanders have suggested that it is where such environments are unpredictable that adaptation to the extreme conditions becomes difficult, Craik suggests that it is because extreme environments are usually small in extent that they support low diversity. He notes (Craik, 1989) that extreme natural environments such as highly acidic or alkaline waters, or hot springs with temperatures above 70°C, are relatively small in extent – and hold few species; by contrast, environments equally extreme, such as cold but unfrozen seas, cover much of the globe and hold a much greater number of species and phyla (Brock and Madigan, 1988). The combination of extremes of conditions and small scale would reduce the chances of colonization by living organisms in the first place and further restrict their evolutionary potential – with a reduced potential for gene flow into, within and between such small isolated areas; Craik suggests that had they been oceanic in extent, hot springs, hypersaline lakes and extreme acid or alkaline waters might resemble the oceans in their diversity.

8.4 SPECIES-ABUNDANCE RELATIONSHIPS

The diversity of a community is affected not only by the actual species composition and species richness; it is as much a function of the relative numbers of organisms in the different species-populations. Such distribution is clearly a consequence of the partitioning of available resources within the community (Brown's 'allocation' rules). The interactions determining resource division within the community will affect not just the number of species supported but their relative population size – with the inherent 'divisibility' of resources determining the potential species number, while the biotic interactions which actually effect that division of resources will themselves determine relative population size. While we have considered in Chapters 4, 5 and 6 something of the detailed dynamics of such resource partitioning, what may be the actual implications for structure within the community as a whole? It is apparent that community structure and dynamics may be affected by the relative numbers of individuals in identified key species known to have a specific impact. Can we also explain more general patterns of distribution of individuals between species – irrespective of what those particular species may be – and what may be the implications of such pattern for community dynamics and stability?

It has regularly been observed over the years, almost as a curiosity, that natural communities contain many more species of rare or intermediate abundance than they do species with high population densities. That is, although in most natural communities there is a relatively large number of species which are represented within the community at low or medium abundance, there are relatively fewer species present at high levels of

Species-abundance relationships

abundance. If this should prove a biological imposition, rather than a mere artefact of sampling or random assortment, analysis of such species-abundance distributions might reveal something fundamental about the basic principles and constraints of community design. If such distributions were, in addition, seen to differ between different communities, they might prove a useful and measurable reflector of some essential difference in the style and operation of those communities. Such observations have led to a number of detailed analyses of species-abundance distributions of both real and idealized communities. Thus some of the distributions proposed are derived simply as empirical fits to observed data (with no explanation offered as to why the species-abundance patterns might show such a form); others are derived from theoretical considerations about how the abundances of species within a community should relate to one another.

Empirical observations and simple models

The first suggestions that communities might show some restricted range of form in the relative abundance of their component populations were presented by Williams (1953), who noted that the frequency distribution of different species of night-flying moths attracted to a mercury vapour light-trap, conformed to the distribution shown in Figure 8.5. To these empirical data, and other data of similar type (e.g. Preston, 1948), have variously been fitted simple logarithmic or lognormal curves (Fisher, Corbet and Williams, 1943; Preston, 1948; Williams, 1953). Both mathematical distributions fit the observed data reasonably well – and can also be extended to describe the abundance curves of other communities or species assemblages, most of which show remarkably similar patterns of relative abundance. But such a fit is not necessarily very informative; we do not know *why* communities should conform to such a distribution.

A different approach was therefore adopted by a number of authors in trying to tackle the problem the other way around. If we make various simple assumptions about the 'rules' by which resources might be partitioned between community members from what we already know of the possible dynamics of such resource use relationships (Chapter 4), we may extrapolate from these initial premises the species-abundance distributions which would result were these conditions met; the observed fit of that derived abundance distribution to empirical data may then be used to justify or refute the assumptions about community design on which the original models were based. In such a way, a more 'reasoned' fit of some mathematical distribution to community data might be derived.

There have been a number of such attempts to 'second-guess' species-abundance distributions. MacArthur (1957) developed the first such model, investigating what would be the observed species-abundance

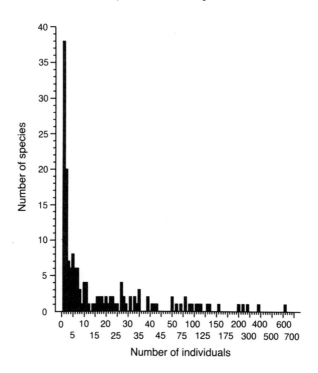

Figure 8.5 The frequency distribution of the abundance of various species of night-flying moths attracted to a light-trap at the University of Southampton's Experimental Station at Chilworth, England in 1992. *Unpublished data Nick Smith.*

distribution were resources within a community partitioned randomly between the component species. Affectionately dubbed the 'broken-stick' distribution because it imagines resources as a stick broken arbitrarily into a number of uneven segments, the model considers what would happen were resources divided purely at random between species, with s species dividing the resources available into s non-overlapping niches of randomly-allocated size. Such a model in fact predicts a species-abundance distribution as illustrated by Figure 8.6. (The figure adopts the more conventional form of presenting species-abundance distributions in displaying the number of individuals in each species as the proportion of the total number of individuals recorded overall (y-axis), with the different species concerned arranged in rank order along the x-axis from the commonest to the rarest; for orientation, the equivalent translation for logarithmic curves, such as that of Figure 8.5, and for lognormal curves are shown on the same axes; MacArthur's broken-stick distribution appears somewhere intermediate between logarithmic and lognormal.)

Species-abundance relationships

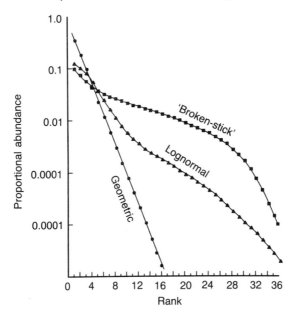

Figure 8.6 Idealized species abundance curves derived from logarithmic, lognormal and 'broken stick' distributions. In this figure, the proportional abundance of each species (number of individuals in that species as a proportion of the total number of individuals sampled overall) is plotted for each species in rank order. (See text, page 128.)

MacArthur concluded that if the underlying picture is one of intrinsically even division of environmental resources, the statistical outcome in terms of species-abundance relationships will be the broken-stick distribution.

The model has its limitations. It assumes implicitly that the environment is dominated by some single key resource to be shared and that this governing resource is to be shared randomly. More seriously, the same distribution may be derived from totally different initial premises; Cohen (1968), for example, has derived an identical distribution of species abundance from underlying assumptions about community structure diametrically opposed to those postulated in MacArthur's simulation, with a multidimensional resource partitioned between the species allowing substantial overlap. In Cohen's models, s species occupy s subdivisions of some multidimensional resource space in such a way that the rarest species occupies one subdivision, the second rarest two subdivisions, and so on, with the most common species occupying all subdivisions; this set of hypotheses leads to exactly the same broken-stick distribution of relative species abundance. Thus, the observation of a broken-stick distribution in the real world does not validate the very specific model initially proposed by MacArthur – or indeed any of the many alternative models which have been developed which predict the same observed distribution of relative abundance.

An alternative set of models substitute some uneven, competitive allocation of resources. The niche pre-emption model of Poole (1974), after Motomura (1932), envisages a partitioning of resources amongst the members of some species assemblage such that the percentage of the total available resources used by each species is determined by that species' success in pre-empting for its own use an exclusive portion of the available resource. The most successful species might commandeer a fraction k of the available resources; less successful species would apportion amongst themselves the resources left in the same manner. (In the original formulation, analysis is restricted to the very special case where the second species is considered to sequester the same fraction, k, of the resources remaining, the third species to occupy a k-th fraction again of the residue, and so on; we may generalize from this extreme case.) If resources are apportioned in this way, the percent of the total number of individuals in each species, ranked from commonest to rarest, presents the familiar logarithmic curve of species abundance.

Once again, however, the assumptions are oversimplified; once again, the same conclusions can be reached from different initial assumptions. Indeed, while a whole host of such models have been developed since the original formulations of MacArthur (1957, 1960), in practice this analysis of species-abundance relationships has not proved as informative as might have been hoped. Poole (1974) concluded: 'Species abundance relationships have rightly been called answers to which questions have not yet been found'.

A provisional synthesis
The fact that empirical data can generally be fitted to a logarithmic or lognormal distribution is striking – but a biological interpretation of why this should be the case is obscure. Nor have attempts to approach the problem from the other end – by making assumptions about the rules of community organization and extrapolating the species-abundance distributions that would result – proved of any great value. Commonly two different distributions, derived from conflicting initial premises, both adequately fit an observed set of data such that even if a hypothesized distribution does fit the observed species-abundance distribution, that fit neither proves nor disproves the postulates of the model.

Perhaps the best synthesis of the whole area is that offered by May (1975b). May notes that the theoretical models resulting in broken-stick, geometric or logarithmic distributions of species abundance do all have certain features in common – in that all consider relatively simple systems, with partitioning of some single key resource considered to be the dominating ecological factor in the determination of community structure; broken-stick distributions result from models which invoke a uniform random partitioning of that fundamental resource, while logarithmic or geometric distributions are characteristic of those models which assume an uneven distribution of resources such as might result from competition. May notes that all such distributions reflect some basic dynamic aspect of the community and will be characteristic in the real world of relatively simple communities (or taxonomic assemblies, like night-flying moths!) whose dynamics are indeed dominated by some single factor.

By contrast, from those models where a number of separate ecological factors are considered, a lognormal relationship is to be expected. Where the distribution of abundance is liable to be influenced by many, more or less independent, factors it will take a lognormal form. This tendency is a function quite generally of all multiplicative processes. MacArthur (1960) pointed out that it is in the nature of the equations of population dynamics that all factors influencing population size shall have multiplicative effects. Where the effects of many independent processes are assembled in such a multiplicative way, a lognormal distribution will result as a simple consequence of a statistical law of large numbers: the Central Limit Theorem of MacArthur (1960).

May concludes, therefore, that for real communities, a logarithmic distribution of species abundance may be expected amongst relatively small sets of species where competition may result in an uneven distribution of some major resource; while amongst small sets of species which randomly apportion among themselves available resources, MacArthur's broken-stick distribution may be expected. For large or heterogeneous assemblages of species, such as whole communities, a lognormal pattern of relative abundance will be expected, as a necessary mathematical reflection of the Central Limit Theorem – although recognition that such lognormal distribution need not necessarily have a biological basis does not necessarily imply that it does not (Box 5).

BOX 5 LOGNORMAL CURVES: BIOLOGICAL IMPOSITION OR STATISTICAL ARTEFACT?

Preston (1962) noted that the parameters of the lognormal distributions describing species-abundance patterns in real communities were restricted in a very particular manner. Lognormal distributions derived from biological data belonged to a rather limited family of 'canonical curves', defined by the fact that the location of the peak of the abundance distribution for the total number of individuals ($N \times S(N)$) coincides precisely with the position of the most abundant species $(S(M)_{max})$ (or $R_N/R_{max} = 1$, where R_N denotes the mode of the plot of individual abundance; and R_{max} is the number of individuals in the most abundant species). Sugihara (1980) claims that such canonical distributions are such a limited subfamily of the full range of lognormal curves theoretically possible, that the regularity with which patterns of abundance in real-world assemblies conform to the restricted conditions of such distribution must have some underlying biological basis. Since the assumption of multiplicative effects does not explain the canonical form of such distributions, he argued, explanations for observed species-abundance relationships must involve ecological and evolutionary considerations. However, in an elegant extension of such analysis, Ugland and Gray (1982) suggest that canonical form – or approximation to its necessary conditions – is rather more commonplace amongst lognormal curves in general than might be expected; they further note that empirical data assembled by Sugihara, while suggesting that abundance curves do approach canonical form for species sets above 200, do not necessarily approximate to such a canonical distribution for species arrays of below 200. Indeed they suggest that the data presented by Sugihara, far from supporting his contention that biologically-based lognormal curves of species abundance are strictly canonical, refute it. For a small number of species, many forms of lognormal (or other) curves may represent the communities; the canonical hypothesis is approximately true only for species-rich communities (Ugland and Gray, 1982).

8.5 THE SIGNIFICANCE OF PATTERNS IN RELATIVE ABUNDANCE

None of these analyses actually tells us why any community shows a particular pattern of species abundance. Stenseth (1979) has suggested that the shape of the species-abundance curve is related in some way to environmental circumstance – being lognormal for communities of stable environments, logarithmic for communities of unstable systems. Evidence for this may be adduced from Patrick's (1963) investigations of the species-abundance relationships amongst diatoms of clear and polluted streams. Results showed clearly that polluted streams supported fewer species and displayed a far less even distribution of individuals between those species – with a few species tending to be extremely abundant. Further support for such a thesis may be drawn from observations of Johns (1984, 1986) on the structure of the avifauna of hill forest in West Malaysia before and after logging. Figure 8.7 shows the species-abundance curves that may be calculated for Johns's data for primary forest, and similar forest six years after selective logging of commercial hardwoods. Fitted curves are calculated from a perfect lognormal and a perfect logarithmic distribution and it is apparent that the pattern of

Significance of patterns in relative abundance 133

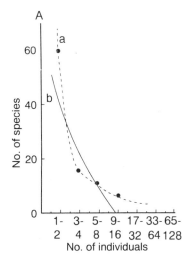
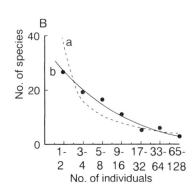

A. Coefficients of correlation:
curve a : r=0.99
curve b : r=0.89

B. Coefficients of correlation:
curve a : r=0.58
curve b : r=0.98

Figure 8.7 Species-abundance curves derived for the avifauna of (A) primary hillforest in West Malaysia, (B) similar forest logged six years before the survey took place. Fitted curves (a) and (b) are calculated from a perfect lognormal distribution (a) and a logarithmic distribution (b). The goodness of fit of each to the observed data points is represented in each case by the correlation coefficient, r, below each graph. Clearly, the pattern of relative abundance of birds of the undisturbed community of primary forest is best fitted by a lognormal curve, while following perturbation the bird assembly better fits a logarithmic distribution. (Unpublished data summarized in Johns, 1987.)

species abundance in primary forest best approximates to lognormal, while that of the disturbed area, even six years after logging, is still best fitted by a logarithmic curve.

Ugland and Gray (1982) have derived lognormal distributions of species-abundance from compartmentation of interactions within communities into discrete cells (Chapters 3 and 4) and conclude that lognormal distributions are indeed characteristic in the real world of systems **at equilibrium**. Such resolution is perhaps more general than Stenseth's rather specific consideration of disturbed or undisturbed communities; after all, both stressed and unstressed communities may be at equilibrium if in the former case the perturbation is maintained. Such conclusions, whether the more specific analyses of Stenseth or the more general inference of Ugland and Gray, are also perfectly in keeping with May's conclusions above (May, 1975b); non-equilibrium systems are liable to be influenced by a few dominant biotic factors, while communities at equilibrium are affected equally by a multiplicity of factors. And all such conclusions do suggest that the species-abundance distribution characteristic of any given community does reflect some underlying characteristic of that community – although it may merely be a passive indicator of the way in which it is ordered, rather than a positive constraint upon organization in itself.

While we do not know why the species-abundance relationships develop in the way they do, we do at least know what they reflect. We may conclude that when one encounters in the natural world a community or assemblage whose species-abundance curve conforms to a logarithmic or broken-stick distribution, then it must reflect a community primarily ordered in respect to one dominating ecological factor (with broken-stick distributions the statistical expression of an ideally uniform pattern of distribution, logarithmic models reflecting some uneven division of resources as would result, for example, from competition). Where one encounters lognormal distribution of relative abundance, one may conclude that this represents a more complex community, ordered by a multiplicity of interactions (May, 1975b).

9
Stability

One of the most important features of any ecological system is its inherent stability – or lack of it. Understanding of the stability properties of any system, and the characteristics of structure and dynamics which enhance or jeopardize that stability, is fundamental to our comprehension of the operation of natural systems and to appreciation of the likely consequences of human intervention in such systems through planned or unplanned manipulations.

Stability is a complex phenomenon influenced simultaneously by many population or community processes – many of which may interact. Many published analyses of stability in ecology have highlighted the effects on stability of one or another specific individual characteristic of population dynamics or community organization in isolation. My aim in this concluding chapter is to review some of those many different 'unitary' relationships which have been established, in an attempt to develop some overview within which the individual relationships between structure and stability may be visualized within their fuller context. To embark on such a review at all is a foolhardy venture: such overview is necessarily oversimplistic and naive, and while I shall attempt to include in that review as many as possible of the various relationships suggested over the years, I shall indubitably have overlooked some and misinterpreted others; nonetheless, such synthesis may prove a useful exercise in encouraging the reader to do better!

Further, in marshalling here the relevant material to focus on questions of stability we may draw together into context the various separate strands left hanging at the ends of earlier chapters, highlighting interrelationships between the different principles established, reconciling apparent contradictions to weave them together into a single whole.

9.1 SOME DEFINITIONS

Ecological stability is variously defined, but may be summarized as the dynamic equilibrium of population, community or ecosystem size and structure. MacArthur (1955) defined stability as 'the ability of both populations and communities to withstand environmental perturbation, to accommodate change'; more formally it is the extent to which the variation in some characteristic of an ecological system is less than the variation in those environmental variables which influence that characteristic (MacArthur, 1957). But from the outset we should stress that stability is not a simple characteristic at all, but a multiplicity of distinct attributes – which do not necessarily respond to changes in characteristics of community structure or dynamics in the same way. Much of the confusion which has bedevilled a very contradictory literature in the past has arisen from a failure to compare like with like, a failure to define terms and concepts sufficiently clearly (see also Pimm, 1984).

Orians (1975) has identified a number of different elements which may be recognized within the overall concept of stability; his rather complex list of stability 'functions' (following Lewontin, 1969) may be resolved into three distinct and quite different 'types' of stability and a number of attributes which may be associated with these (Table 9.1). The three basic types of stability itself (perhaps better regarded merely as different facets of the same phenomenon) may be considered as **constancy, resilience** and **inertia**. By **constancy** we refer to a lack of change in some parameter of a system, such as number of species, taxonomic composition, life-form structure of a community, size of a population or some feature of the environment. **Resilience** may be considered as the ability of a system to recover and continue functioning even though it may have changed its form. Thus a community might be described as resilient if, during or after some disturbance, even though its composition and structure may have changed substantially, it is able to continue functioning as a viable system. Finally, **inertia** is the ability of a system to withstand or resist such perturbation in the first place; strictly speaking it is to such inertia that MacArthur's (1957) definition refers.

Attributes of such stability functions are Orians' **persistence, elasticity** and **amplitude**. Thus elasticity is a measure of the speed with which a system returns to its former state following some perturbation; (this elasticity equates to what Pimm and Lawton (1980 et seq.) and Pimm (1982, 1984) mean by their use of the term 'resilience' (here defined rather differently) and with what is measured by their 'return time'.) The **amplitude** of a system defines that domain over which it is stable; a system has high amplitude if it can be considerably displaced from its previous state and still return to it. Where a system can return to its previous state following any perturbation, however large, it is con-

Some definitions

Table 9.1 Stability functions: concepts and terms used in discussions about stability. (Adapted from Orians, 1975, after Lewontin, 1969.)

Term	Concept
Constancy	A lack of change in some parameter of a system.
Inertia	The ability of a system to resist external perturbations.
Resilience	The ability to continue functioning after perturbation.
Persistence	The survival time of a system or some component of it.
Elasticity	The speed with which the system returns to its former state following a perturbation.
Amplitude	The area over which a system is stable.
Cyclical stability	The property of a system to oscillate around some central point or zone.
Trajectory stability	The property of a system to move towards some final end point or zone despite differences in starting point.

Orians himself notes: 'This listing of the meanings attached to the concept of stability is not intended as a classification system because the terms are not comparable. Constancy and persistence are descriptive terms implying nothing about underlying dynamics. Cyclic and trajectory stability have measures of inertia, elasticity and amplitude associated with them, etc. The separation of concepts is presented only to illustrate the many meanings of stability, the existence of which presumably reflects a need for a variety of notions relating to fluctuations' (Orians, 1975).

sidered 'globally stable'; where such return is recorded only after relatively minor displacement it may only be accorded 'local stability'. **Persistence** refers to the survival time of the system or of some defined element within it. In this sense, one population might be considered more 'stable' than another, if its mean time to extinction were longer (Roff, 1974). Finally, Orians' concepts of cyclical or trajectory stability merely provide further descriptors of the form of stability observed, in relation to whether the system tends towards an end-point of a single equilibrium point, or to stable 'limit cycles' (May, 1972b, 1976).

Orians's terminology will be employed throughout this chapter, in preference to terms proposed by other, later authors (e.g. Harrison, 1979; Pimm 1979, 1982; Pimm and Lawton, 1980; Diamond and Case, 1986), as more precisely separating functionally-different aspects of stability – and in having the authority of precedence. The reader should be aware, however, that the use of the term resilience here to denote some intrinsic capacity to continue to function in some form, differs from Pimm and Lawton's definitions of resilience as a measure of elasticity.

9.2 STABILITY OF COMMUNITIES

While the dynamical stability of ecological communities is in part conferred upon them by the inherent stability of their component species populations, other emergent characteristics of the community may independently show additional stability properties in their own right. Thus we may attribute some stability in structure of the community: a constancy of trophic form of food web design, a resilience in the mode and balance of the community's operation; it is unclear whether constancy in other structural characteristics, such as species-abundance distributions are cause or consequence. What is the evidence for such stability?

Factors affecting the stability of individual species populations and interacting pairs of populations have been widely reviewed by, for example, May (1975d, 1976, 1981), Hassell, Lawton and May (1976), Begon and Mortimer (1981, 1992) and many others; such analyses will not be reviewed in detail here. It is apparent that any community can only be as stable as its constituent populations. The relationship is indeed two-directional: interactions between the populations of any community have a powerful influence on the dynamics and potential stability of those populations themselves; equally the stability of populations that results contributes to the stability of the whole community of which they are a part.

Yet it is clear that it is the *interactions* between its component populations which may influence community stability. And the influence of one population on another – or the effects of perturbations in population size of one upon the population size of the other – depends heavily on the nature of the interaction and whether such relationships are donor-controlled or more powerfully influenced by population changes among the 'recipient'. The interactions between populations are neither unidirectional nor necessarily of equal effect in the two directions. Armstrong (1982), for example, makes it clear that the interaction strength of a single link between two organisms within a community web may have markedly different magnitude and different consequences for stability in the two directions. In addition, the type of link involved (predatory, competitive, mutualistic, etc.) assumes considerable importance, as does the topology of the entire web of interaction. Thus observed communities are in general more stable than randomly constructed systems with the same number of species because of non-random organization of the links formed (Lawlor, 1980; Sugihara, 1984).

The illustration perhaps most widely-quoted in support of apparent stability properties at the level of the entire community is the stability of trophic structure reported by Heatwole and Levins (1972), after Simberloff and Wilson (1969), in their analysis of the organization and structure of arthropod communities established on mangrove islets in Florida

Stability of communities

Keys following defaunation (Chapter 1). Their results appeared to show striking constancy in trophic structure of such mangrove communities; that despite the fact that the detailed taxonomic composition on any island following recolonization was usually altogether different from that observed before defaunation, the total number of species in any instance and the distribution across trophic categories appeared to be very much as before.

In practice, such evidence is less convincing than it might at first appear. We have already noted Simberloff's assertion that the structure of the communities formed after recolonization cannot be shown to be more similar to previous structures than that which would have resulted from random colonization (Simberloff, 1976). Further, the data quoted by Heatwole and Levins in any case chronicle the re-establishment of new communities following total extirpation of the previous fauna; they do not report the recovery of a disturbed system. The fact that the new mangrove communities that develop share in any instance many features of the old (in terms of species number and trophic design) cannot be taken as any evidence of constancy or resilience of structure of the old community. Rather the results are better considered as pointing towards the same 'fixity of design' under given circumstances explored on page 80; that there is, so to speak, some optimum structure within a community of given function that utilizes available resources most efficiently. Evidence for the actual resilience or constancy of an existing community can only come from studies where the community has been disturbed but not entirely destroyed.

The fact that the secondary succession in ecological communities generated after some perturbation eventually returns to its previous climax state – although commonly through different developmental pathways – may be adduced as evidence for resilience, and for the constancy of such climax-communities (although we may have reservations about the interpretation in this instance too: page 100). A more specific example may be drawn from the recent analyses by Johns (1984, 1986) of faunal assemblies in montane forest in West Malaysia before and after selective logging. Readers of a previous book, (Putman and Wratten, 1984) will be familiar with the details of this analysis. Johns conducted a complete census of the vertebrate faunas of hill forest areas of West Malaysia, surveying areas of primary forest before logging operations commenced, during logging and up to six years after extraction of timber had ceased. Logging was selective for main timber species only and at a rate of only some 18 trees per hectare for the study area as a whole. This may thus be classified as perturbation – not destruction – of the existing community.

The immediate effects of logging are clear, with the numbers of species of both birds and mammals recorded declining dramatically – and with a gross shift in species composition of the whole faunal assemblage. Many resident species disappeared altogether from logged forest, but other new species arrived within logged areas to exploit the regeneration; thus of 20 species of mammals recorded in forest areas logged one to two years previously, only 15 remain of the 45 members of the original community and at least five species were entire newcomers. Even among those native species that remained Johns recorded marked changes in dominance; diversity of bird faunas fell not solely as a consequence of a decrease in species number but due to marked changes in equitability. After five to six years logged areas had begun to show evidence of recovery. Trophic structure of the mammal fauna had returned towards its original state (Table 9.2); species composition and species number have changed, but the relative balance of numbers in different trophic categories suggests some recovery towards the original structure. Similar changes are recorded in diversity and trophic organization of the birds (Johns, 1986). Taken as a whole, Johns' careful analyses suggest that the communities studied show little inertia (community structure was substantially affected even by relatively slight perturbation), low constancy (some recovery of trophic structure but no constancy of species composition), but reasonable resilience.

What factors, then, may confer these separate properties of constancy, resilience and inertia on natural or man-made communities?

Table 9.2 Trophic strategies of mammals of West Malaysian hill-forest before logging and following a period of 5–6 years recovery after logging. The numbers of species (excluding Chiroptera, Muridae) in each trophic category are accompanied, in brackets, by corresponding percentages of the total sample. (After Johns, 1983.)

Trophic categories	Primary forest		Logged forest after 5–6 years	
Terrestrial folivore	(12)	5	7	(25)
Arboreal frugivore/folivore	(5)	2	2	(7)
Terrestrial frugivore/folivore	(5)	2	1	(4)
Arboreal frugivore	(19)	8	6	(21)
Terrestrial frugivore	(2)	1	0	(0)
Arboreal insectivore/frugivore	(21)	9	6	(21)
Terrestrial insectivore/frugivore	(10)	4	2	(7)
Arboreal insectivore	(2)	1	0	(0)
Terrestrial insectivore	(5)	2	1	(4)
Predators	(19)	8	3	(11)
		42	28	

9.3 EFFECTS OF POPULATION INTERACTION

In so far as the stability of any community is contributed to by the stability of its component populations we should perhaps review briefly the general consequences of different types of interaction for *population* stability. We have already established (Chapter 2) that interspecific competition and predation have in general a destabilizing rather than a stabilizing influence on population dynamics. While both types of interaction may, so to speak, be 'tolerated' by a population, May (1976) has characterized predator–prey, or parasite–host systems as in some tension between the stabilizing effects of prey density-dependence and the often destabilizing predator functional and numerical responses. Similar conclusions can be drawn for competitive interactions in balance between the stabilizing effects of intraspecific interaction and the destabilizing effects of interspecific competition.

Few interactions are, however, symmetrical in their effects. Amongst the various forms of population interaction that have been modelled we should draw a clear distinction between 'top-down' relationships, where the dynamics of the population on the receiving end of the interaction are more strongly influenced by that interaction, and those donor-controlled interactions where the dynamics of the 'prey' are relatively independent of the effects of the interaction but those of the 'predator' are strongly influenced by what happens to its prey population. Top-down interactions as a rule are far more likely to be destabilizing for populations of both interactors, while interactions where the predator's population merely tracks that of its prey will in general be far less disruptive. The destabilizing effects of both predator–prey and competitive interactions may further be mitigated by the effects of context: wider interactions such as those of diffuse competition and competitive mutualism may help improve the dynamical stability of interacting populations – while complexity or heterogeneity of the physical as well as the biotic environment may further facilitate stable interaction (pages 31–2).

The potential stability of simple systems of interacting populations has been reviewed in graphical form by Jeffries (1974), who presents a simple set of 'rules' to be observed if such interactive arrays are to be certain of showing dynamical stability. Jeffries offers a simple test for assessment of the likely stability of simple systems of interacting populations, at least in regard to that element of stability contributed by their dynamic properties (Box 6).

BOX 6: QUALITATIVE STABILITY AND JEFFRIES' 'COLOUR TEST'

Any ecological system is considered 'qualitatively stable' when one may infer stability simply on the basis of the types of interaction between its member species (with no consideration of their magnitude). As described in Chapter 3, any system of interacting organisms can be represented mathematically quite simply as an n by n matrix summarizing both sign and magnitude of the interaction coefficients of each pairwise combination of species; conventionally the coefficients used to construct such a matrix are the Lotka–Volterra coefficients of interaction: $\alpha_{i,j}$. Such community matrices indeed form the whole basis for analysis of the dynamics of multi-species food webs and model ecosystems (e.g. May, 1973a, 1975d *et seq.*).

Suppose that a given community, based on some given matrix, is found to show local stability in returning to its former state following minor perturbation. Suppose further that the magnitudes (but not the signs) of the interaction coefficients within the matrix can be interchanged at random, and yet each new matrix so derived shows a similar capacity to return to equilibrium. In such a case the stability of the system depends only on the qualitative nature – that is, the sign – of the non-zero entries of the interaction matrix and not the strength of any interaction. Clearly, community matrices which are qualitatively stable in this way form only a robust subset of the entire array of potentially stable matrices and recognition of such qualitatively stable systems does not provide a comprehensive method of identifying all stable or unstable combinations; none the less, if we are able readily to identify such systems, we can at least distinguish between those community structures which will definitely show local stability over all parameter values and those which are less likely to be stable, or stable only under very specific conditions.

Jeffries represents the various interactions summarized within a community matrix in a series of 'digraphs' (after Roberts, 1971) where the different organisms are linked to each other by a series of connecting, directed lines. Thus the simple digraph of Figure 9.1a shows that the size of the population of species 1 affects the rate of change both of its own population (through some intrinsic, density-dependent feedback mechanism) and that of species 2; the population size of species 2 does not however exert any influence on future change in species 1. The signs written near to the connecting lines indicate the qualitative nature of the effect of one species upon the other and correspond to the signs of interaction in the community matrix. In Figure 9.1b, species 1 has a negative effect on species 2, while population size of species 2 has a positive influence on future population size of species 1; the figure represents perhaps a parasitic relationship

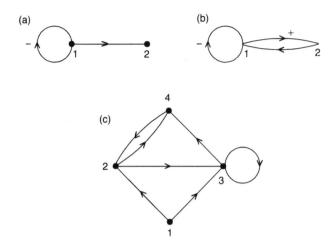

Figure 9.1 Simple digraphs representing population interaction. (a) Species 1 is self-regulatory, with negative feedback on population growth from its own population size; the size of population of species 1 also impinges on species 2, but there is no reciprocal effect of species 2 on species 1. (b) Species 1 is self-regulatory; but here, population size of species 1 has a positive effect on that of species 2, while species 2 exerts a negative impact on populations of species 1 (a predation loop). (c) A digraph with a 1-cycle, a 2-cycle and a 3-cycle. (After Jeffries, 1974.)

between parasite (species 1) and host (species 2) or a predator–prey system. Many of the interactions within a more complex digraph (such as Figure 9.1c) are cyclic: such p-cycles are defined as a set of p points, (species), connected together such that a circuit may be traced between them following p directed lines. (Each p-cycle must involve precisely p lines – thus figures of eight and so on are not true cycles.)

From such diagrammatic representations of simple communities and their constituent species-interactions, Jeffries develops a simple test for qualitative stability (extending upon more mathematical treatments of May, 1973b). Simply, individual 'species' (points within the digraph) are coloured black if they are self-regulating (as for example, species 1 in Figure 9.1a). Each predation loop, or subsystem within the entire community digraph, is then assessed to see if the various points can be coloured in such a way as to satisfy simultaneously four criteria:

1. As noted, each self-regulating point is black;
2. The digraph, or predation loop contains at least one white point;
3. Each white point within the system is connected by a predation link (in practice any +/− link) to at least one other white point;

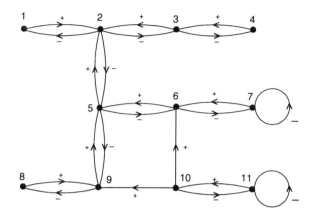

Figure 9.2 A more complex digraph with two predation loops or subsystems. Species 1 to 9 comprise a discrete subsystem within the entire community digraph, as do species 10 and 11. According to Jeffries' colour-rules (page 143), species 1, 2, 3, 4, 8 and 9 may be coloured white and species 5, 6 and 7 coloured black; this system thus passes the colour test – and is not necessarily stable. By contrast in subsystem 10, 11, species 11, being self-regulatory must be coloured black – and by rule (4) species 10 must be black as well; this subsystem thus fails the colour test.

4. Each black point connected by a predation link to one white point is connected by a predation link to at least one other white point.

If it is possible so to colour-code individual species points, the system is said to satisfy Jeffries' colour test (Figure 9.2).

The development of this argument may seem somewhat oblique – and the reader might be forgiven at this point for questioning the relevance of all this to stability. However, having got thus far we may now assess any community matrix for qualitative stability on the basis simply of the signs of interaction. Jeffries shows that for a community to show qualitative stability:

1. All species within the matrix must either show no density-dependence or some degree of negative feedback in population growth; for stability no species may show positive feedback (more formally, after May, (1973b), $\alpha_{i,i} \leq 0$).
2. Every pairwise interaction between species must contain one positive and one negative interaction coefficient ($\alpha_{i,j} \times \alpha_{j,i} \leq 0$ for all pairwise interactions). Competitive (double-negative) and mutualistic (double-positive) interactions are destabilizing and

automatically compromise the universality of qualitative stability.
3. Digraphs can contain no p-cycles with $p \geqslant 3$.
4. All subunits (predation loops) within the community digraph must fail the colour test: that it is impossible to colour-code the different species points within the digraph to satisfy all the conditions of the colour test.
5. There must be at least one way in which one may erase all but n connecting lines in the community digraph, leaving each of n species still within a p-cycle.

Jeffries' analyses are elegant in their simplicity – and while, as we have noted, such a 'colour test' cannot categorically demonstrate that a given community matrix is necessarily unstable (some may still show strong stability over a specified range of parameter values), it does prove an effective rule of thumb for identifying at least those community arrays which are stable over all parameter values. At the same time, analysis also offers a number of important insights in more general terms into the effects on stability of population interactions, when these are considered within the context of multiple interactions within a multi-species system, highlighting the required topological structure of the interactions within the community web and showing precisely how the various different types of interaction may be arranged within the matrix if the community is to display such robust qualitative stability.

9.4 DIVERSITY AND STABILITY

The fact that the destabilizing effects of primary interactions such as predation or interspecific competition may be smoothed away within the context of a more complex matrix of interaction has frequently led to the supposition that in large measure, some of the dynamical stability of ecological communities is conferred upon them by their very complexity. The observation by authors such as Elton (1958), Odum and others that the most 'stable' communities (tropical communities by comparison to temperate or arctic ones, late successional communities by comparison to pioneer stages; the measure of stability is not defined) tend to have a greater complexity of structure and greater species diversity, led to uncritical acceptance of a general axiom that diversity begets stability.

Intuitively such a relationship seems reasonable. After all, the destabilizing effects of competition may be reduced by indirect interaction within a multi-species array – the effects of a third competitor or a major predator. The injurious effects of predation upon the dynamics of the prey may be ameliorated if the predator has alternative prey species to which it may switch its attention if one prey species suffers a substantial decline (whether as a result of that predation or through some other independent event). Irregularities within a complex system are easily compensated for by minor adjustments elsewhere. Nor is the proposed link between complexity and stability built entirely upon such armchair reasoning, or anecdotal observation. The relationship has apparently been established by a number of early experimenters (Pimentel, 1961; Odum, Barrett and Pulliam cited in Odum, 1971; and others).

But regrettably, just because something is intuitively reasonable does not necessarily mean that it is correct. Is stability the result of the observed diversity, or is the diversity a consequence of the inherent stability? May (1972a, 1973a, *et seq.*) noted a tendency for model communities to become *less* dynamically stable with increasing species number rather than more stable; his analyses suggested that increasing species number or connectance within randomly-assembled webs decreased the stability of constituent species populations. From such analyses May concluded that complex, dynamically fragile communities can develop only in predictable environments where the system need cope with only relatively small perturbations (and thus where, if perturbations remain small, its intrinsic lack of stability is of lesser moment). By contrast, in an unpredictable environment, there is need for the region over which any one population remains stable to be extensive, with the implication that the system must be relatively simple to persist (May, 1981). May concluded that the frequently observed correlation of high diversity with high apparent stability was not in itself a causal relationship, but rather a reflection of two independent consequences of a common cause: that environmental stability permitted the development of high diversity while at the same time ensuring the continued stability of a high-diversity system.

Random assembly is perhaps an unreasonable assumption; subsequent authors have studied more biologically reasonable webs, which relax constraints on structure and assemble webs according to more realistic criteria, but the conclusions remain essentially unchanged: any increase in species richness or connectance within such webs decreases their dynamical stability (DeAngelis, 1975; Lawlor, 1978; Pimm, 1979). May's original conclusions seem incontrovertible, yet so counter-intuitive.

Closer examination, however, allows some reconciliation. To begin with we should note that all such analyses relate to simple constancy: the stable equilibrium of species composition and population size. To some extent, what became known as 'May's Paradox' is resolved in the recognition that his conclusions and those of subsequent analysts thus relate primarily to community constancy; the idea that the more complex is a given system, the greater will be its potential to accommodate change through minor adjustments elsewhere within its structure – in some sense 'redirecting' the community relationships to absorb the perturbation – is a concept, by contrast, that considers the community's resilience. If we thus distinguish the separate facets of stability, May's conclusions no longer conflict with intuition. The more complex a system becomes, the less likely it may be that it is able to maintain constancy of species composition, population sizes, etc. By contrast, the more complex it is the more likely it is to be able to continue functioning in some form following perturbation, because of its capacity to absorb such perturbation, as 'damage' to one part of the system is compensated for by adjustments in other parts of the complex whole.

In such analysis it may also be helpful to distinguish more clearly between the two separate components of complexity: species richness and diversity of interaction. Community resilience is in fact better regarded as a function of the diversity of energy-exchange pathways within the community rather than of simple species number; the two are not necessarily connected in any direct way, for the link between the diversity of energy-exchange routes and species richness will depend on the types of organisms involved and to what degree they are specialists or generalists. In such terms we may perhaps re-examine the relationships between all the various parameters of model, and real, food webs and community stability. The reader is also referred to Pimm (1984) for an alternative analysis of the relationships between community complexity and stability.

9.5 FOOD WEB PROPERTIES: THE INFLUENCE ON STABILITY OF SPECIES RICHNESS, CONNECTANCE AND INTERACTION STRENGTH

The mathematical analyses of May and later authors, reviewed briefly above, suggest that an increase in complexity within a matrix of interaction leads to a reduction in the ability to maintain **constancy**: that increase in species richness or connectance reduces the potential of a system to return to its former equilibrium structure following perturbation. Couched in this more restricted form we will not dispute such a relationship, but there are a number of additional points we should perhaps note. Conclusions from the study of model community matrices that complexity decreases the inherent stability of structure and dynamics are all based on mathematical analyses of local stability properties of simple linearized matrices. Local stability is defined as a return to equilibrium following an arbitrarily small perturbation: May's models do not assess relative stability in relation to major perturbation, or investigate the true level of global stability (quantifying amplitude) – and further seek a **very exact return to precisely the former state**. As noted by Hastings (1988), perturbations in the real world are commonly not small; often the really relevant biological question is one of persistence (a persistent system may be loosely defined as one in which no species actually go extinct) and local stability is not a necessary condition for persistence (Hastings 1988). These arguments are developed further by Law and Blackford (1992).

Further, while a decrease in local stability with increased species richness or connectance is observed in the majority of model systems studied, these have for the most part been modelled according to top-down Lotka–Volterra dynamics; increased complexity is not associated with any necessary decrease in stability in models with strongly donor-controlled dynamics (Pimm, 1984, and others) and there is a continuing lively debate about the relative importance of top-down or bottom-up relationships in natural communities (page 42, here and see Matson and Hunter, 1992; Hunter and Price, 1992; Menge, 1992). Destabilizing effects of increasing complexity are also not observed where models incorporate strong self-regulation of component populations (Haydon, 1992) or elements of spatial or temporal heterogeneity (above, page 141) – indeed any factor which admits an element of asynchrony in population dynamics. Thus the conclusions are still to an extent dependent on the form of the model used.

Earlier consideration of the relationships between complexity and stability had suggested there might be a positive link between the diversity of energy exchange pathways within a web and the **resilience** of that system. Stability in this sense relates to an ability to accommodate disturbance to some portion of web by shunting material through other routes (MacArthur, 1955). Diversity of energy exchange pathways need not necessarily correlate with an increase in species richness overall; such diversity might indeed be accommodated by an increase in the number of species if the species matrix is composed primarily of specialists, but could equally result, with no change in S, if existing species became more generalist in habit. Recognition, however, that in stable communities species are generally limited in the number of links in which they may engage (to an average of 3–5) and that in consequence, product SC approximates to a constant, suggests that overall diversity of interaction may show relatively little variation between systems. Such formal studies as have investigated the relationship between web connectance and constancy suggest that increasing connectance decreases constancy both of species composition within a web and of individual species abundance, although overall persistence may be increased (e.g. Robinson and Valentine, 1979) and average return time to equilibrium after disturbance may be less in systems of higher interaction (Pimm, 1979, 1984). The further observation by Briand (1983) and Briand and Cohen (1984) that SC is systematically higher for food webs in constant, rather than fluctuating, environments (i.e. for given S, connectance is higher in constant environments than in variable ones) accords well with the idea that the development of complexity is indeed a consequence of ecosystem stability rather than a cause.

The strength of interaction of particular linkages within a community web will clearly have profound implications for population stability of the interactors (above, page 141); average interaction strength within the web as a whole may also influence inertia, resilience or constancy. Because of the observed relationship between interaction strength (b), species richness (S) and connectance (C) (Gardner and Ashby, 1970; May, 1976, 1986a), for any given species richness S, any increase in b should result in a decrease in C (or, since S is constant, a decrease in the observed number of links per species). This will result in a smaller number of links, but each link will be stronger. Given that product SC actually shows considerable variation between communities, however (Cohen and Newman, 1985; May, 1986a), even with given S, b may change without any resultant change in C. What then is the effect of an increase in b with the same number, or fewer observed links?

High interaction strength will result in changes within one part of the community web being rapidly and strongly passed to other parts of the system; the system shows no ability to absorb perturbation (*sensu* MacArthur, 1957) – thus resistance or inertia will decrease. On the other hand, high interaction strength also implies quick action in buffering out the effects of population change amongst closely linked sets of interactors, so should result in a quick return time and high constancy (of species composition and population numbers) as well as high resilience.

Hopkins has presented a useful analogy (in Putman and Wratten, 1984) for visualizing some of this interaction between species richness, connectance and interaction strength in comparing the links within a community web to the arteries or capillaries of the mammalian circulatory system. In such an analogy simple, species-poor systems will be seen to have few links within the web as a whole, but these links will be strong – analogous to arterial connections. Such arteries are robust and although blockage of any one of these restricted pathways for energy exchange would have serious repercussions for the community as a whole, blockage of such robust linkages is in fact highly unlikely. Such arterial systems are thus of high inertia, and in consequence of high constancy, but once perturbed have low resilience. In more complex systems, however, the slender links between individual species populations are, like capillaries, much more easily blocked – but such blockage is inconsequential in terms of continued community function. Such systems with high connectance and low average interaction strength will show high resilience, but low inertia.

9.6 WEB TOPOLOGY AND COMPARTMENTATION

In many communities we may identify apparent compartmentation of structure: links between species do not span the entire species array but are organized into discrete cells that seem tightly interconnected within themselves but only weakly connected with other parts of the community as a whole. Such compartmentation of community webs offers a way of structuring interaction to enhance stability for any given interaction strength averaged over the web as a whole; the implications for stability have been examined by May (1972a, 1973a) who argues that for a given species number and web connectance, model food webs have a higher probability of being stable if the interactions within them are organized in this way into blocks. Food web subcompartments have been identified by Pimm and Lawton (1980), corresponding to discontinuities in distribution of the entire community between different habitats or microhabitats, although their existence more generally in the absence of such structural templates is unproven. Subunits within the community matrix structuring competitive interactions have been clearly identified in recognized guilds (Chapter 5) although the question of whether they, too, reflect structural discontinuities in the distribution of resources, or a more 'active' functional construct is equally debated.

The possible importance of guilds in conferring stability upon the community is championed by Diamond (1975 *et seq*.) who asserts that such guilds are assembled specifically so that their combined resource consumption curve precisely matches the resource production curve of the community, minimizing the unutilized resources available to support potential invaders and thus, effectively, rendering the guild immune to invasion. Such guild structure thus enhances constancy and inertia within the guild itself.

Further characteristics of the web design of stable matrices, believed to have some structural importance in maintaining that stability, are summarized by Pimm (1980a, 1982) as: (1) food chains should be short; (2) species feeding on more than one trophic level (omnivores) should be rare; (3) those species that do feed on more than one trophic level should do so by feeding on species in immediately adjacent trophic levels (Pimm, 1980a). To such a list we might now add a recognition of restrictions on the topological structure of the web – the nature and direction of certain minimal linkages required to provide a mathematically rigid structure (Sugihara, 1984) – and that stable food webs are characteristically 'upper triangular' (Cohen, Newman and Briand, 1985; Cohen, Briand and Newman, 1986). The extent to which such characteristics are merely passive reflections of some deeper underlying constraint, or in themselves do represent dynamical constraints on community structure through a direct influence on the stability properties of its interacting populations remains open to question (Lawton and Warren, 1988; Nee, 1990).

9.7 SUCCESSION AND STABILITY

Inherent stability (in the sense of constancy of structure and composition – in the absence of serious perturbation) increases tautologically through ecological succession; climax communities by definition are determined as the end point of directional change within the community. Accompanying the progressive changes in structure and species composition which occur throughout succession there may also be recognized characteristic changes in the bionomic character of member species and in the dynamics of the community itself. Thus organisms characteristic of (stable) climax systems tend to be larger; life cycles tend to be more complex, with longer generation times; there is a noticeable increase in niche specialization, itself accompanied by an increase in the importance of contest competition. All these different attributes in fact characterize a shift in emphasis of selection pressures imposed on the organisms within the community towards a greater importance of those features which promote a balanced equilibrium with stable features of the environment [K-selection] as opposed to those favouring maximum population growth [r-selection] and such changes in 'style' are equally apparent in analysis of any community approaching equilibrium – as structure and dynamics are more strongly controlled by biotic interactions with other members of the community than by the effects of stochastic variation in environmental conditions (Chapter 7).

Such changes in the attributes of organisms characteristic of equilibrium systems reflect other underlying changes in the dynamics of the entire community. Thus we may record during successional development a progressive increase in the total organic matter of the community; nutrient cycles tend to become closed as more and more nutrients are bound up within organisms rather than being free or extrabiotic and as the system develops increased efficiency of conservation of nutrients; the decomposer element of the community becomes larger and more important and finally, the flow of material around the community becomes progressively slower (Odum, 1969; DeAngelis, 1975).

Clearly a number of these features are interrelated. Increased conservation of nutrients is related to the increase in size and importance of the decomposer level within the community. The fact that nutrients tend to be bound up within organisms and that the amount of 'free' nutrients declines is a function of the progressive increase in organic matter of the community and its enhanced complexity; these last two factors also explain why flow of material around the community becomes far slower, etc. (Putman and Wratten, 1984). Most significantly all these changes during succession reflect a change in the general 'economy' of the community, a change so to speak in policy.

All these separate changes are in fact a direct result of the primary increase in organic matter within the community. The community emphasis shifts from the rapidly growing, rapidly changing and expanding system characteristic of pioneer communities and early seral stages to a more measured, fully exploited system. During the early stages of succession, the total gross primary production within the community (GPP) is far in excess of that required for the respiratory needs of the community at all trophic levels. Since community production exceeds community respiration there is a net surplus of energy/organic matter which may be devoted to growth; the community accumulates biomass. As time goes on and biomass increases within the community, so respiratory demands also increase until the total respiratory energy use by the community exactly balances community production and the whole process must stop. This one feature alone explains all the other changes in resource flow through succession that we have just catalogued; further, such analysis of the changing relationship between community production and community respiration (CP:CR) offers in itself a very much simpler conception of the successional process as a whole as, simply: the accumulation of biomass within a community (Cooke, Beyers and Odum, 1969).

Resource flows alter through succession, but also may differ more generally between ecological communities of different style or dynamics. Can we draw any more general conclusions about the relationships between patterns of energy use and nutrient flow within a community and its stability?

9.8 NUTRIENT DYNAMICS AND ENERGY FLOW

Much of the character of equilibrium systems is determined by their saturation – the fact that resources are fully utilized and that thus the dynamics of the community are dominated by biotic interactions between member species in their use of those resources as community respiration precisely balances community production. Where $CP>CR$ the community has a surplus of production, 'invites' invasion and will continue to change (page 101 here and Diamond, 1975); in our terminology here it will have low constancy or inertia, though clearly resilience is high. More rarely, certain communities may be shown to be hypertrophic, with $CP<CR$; clearly such communities cannot be sustained without subsidy and by definition must suffer some change of state until once again CP and CR are at equilibrium. As an aside we might note that this requirement for energetic balance also offers an explanation for arrested successions: natural or manipulated ecosystems where succession appears to 'stop short' and stable communities are established at some

pre-climax stage or **plagioclimax**. If such communities are to have halted in successional development they must somehow have reached a balance between CP and CR before the true climax – a balance easily achieved and maintained if there is a net export of production, or regular harvest from such ecosystems (for a fuller development of these arguments, see Putman and Wratten, 1984, pp. 353–4).

Nutrient dynamics also show marked variations between communities. We have noted that during succession more and more nutrients become bound up within the biota, that nutrient cycles tend to become more closed; conservation of nutrients is high, but the speed of recycling decreases. What may be the implications of such variation in nutrient dynamics more generally for stability?

We have already noted that seemingly unstable model communities may display a measure of persistence unexpected from their structure if resource flows through that community are rapid (DeAngelis, 1980); the availability, predictability and utility of all resources will clearly also have a profound influence on the resilience of any system. The wider implications of patterns of nutrient flow for stability are treated in more detail by DeAngelis (1992).

9.9 ENVIRONMENTAL CHARACTER

Throughout much of the foregoing we have been repeatedly drawn back to considerations of the influence on stability of environmental character – whether directly or through intervening variables. Thus the constancy and predictability of the environment affects the bionomic characteristics of individual organisms: the extent to which they are physiological specialists, ecological generalists, r-strategists. Environmental character may determine the patterns of resource flow: nutrient cycles and the nature of the energy flows within the system. May's controversial analyses (1972a, 1973a, 1975d) suggest that complex communities are dynamically fragile (at least in regard to a rather restrictive definition of constancy) and that the observed correlation between complexity and stability must derive from the fact that such fragile communities can develop only in predictable environments where the system need cope with relatively minor perturbations: that environmental stability permitted the development of high diversity while at the same time ensuring the stability (constancy) of a high-diversity system. That link between diversity and environmental character may certainly be substantiated; we concluded in Chapter 8 that species abundance distributions strongly reflected the level of environmental disturbance, while species richness appears a direct function of the severity, constancy and predictability of the physical environment (Slobodkin and Sanders, 1969). Clearly, whether through a direct influence on stability, or through its shaping influence on other properties of community structure (themselves potentially increasing or decreasing the intrinsic stability of the system), the characteristics of the physical environment do exercise a very powerful effect on the stability of such communities as ever reach equilibrium.

9.10 CAUSES OF STABILITY IN ECOLOGICAL SYSTEMS

Figure 9.3 presents only some of the many facets of the structure and character of communities that may have some bearing on stability. The diagram is presented in an incomplete form – for none of the links between the various community attributes or environmental factors and the different facets of stability have been drawn in, nor their influence in enhancing or decreasing constancy, resilience or inertia indicated. Clearly, many of the features noted are themselves interrelated, and similar links might be drawn to show the positive and negative influence of these 'independent' variables of environmental character or community structure upon each other, as well as their separate or combined effects upon the stability properties affected. The basic diagram is designed to be photocopied – for in drawing on it for yourself the various possible links, and thinking through again their likely effects in increasing or decreasing stability, such exercise may help clarify in your own mind the interaction and integration of all the different parameters we have discussed.

Even within the restricted schema presented in Figure 9.3, some of the relationships between the different variables, or their effects upon stability, remain unexplored. In many cases the lack of variation in structure of natural communities in regard to some given characteristic precludes formal analysis of its effects on stability – a problem the more frustrating because the very constancy in that parameter may not reflect any necessary lack of variation, but may itself be due to the fact that the majority of natural communities for which it has been measured are (in that they have persisted for long enough to warrant study) by definition stable. In other cases (for example, the implications of food web topology; Sugihara, 1984; Auerbach, 1984) investigations are still in their infancy.

In final conclusion however we must note that while throughout this book we have been searching for patterns in natural communities, seeking explanation for such patterns – and assuming that those patterns once resolved may reflect some necessary constraint of structure and dynamics which enhances stability – this may itself be a false quest. The patterns we observe, far from promoting stability, may simply minimize instability or reflect a derived structure which is in some way resistant to change but has no more fundamental significance.

The dominating influence of environmental fluctuations in determining the level of biotic interaction expressed within a community and their importance in influencing structure in relation to the effects of stochastic variation in environmental conditions; and the clear role that environmental factors continue to play even within equilibrium communities in determining structure and diversity, in fostering the persistence of inherently unstable assemblies, certainly adds weight to May's suggestions that perhaps rather little of the stability of natural communities is derived from their own structure or internal organization, but that complexity of structure, and diversity of both interaction and species, are instead determined and simultaneously constrained by the stability and productivity of the environment itself.

Indeed it has been widely proposed that far from being inherently stable, or 'seeking' stability of structure as we have just suggested, most climax communities are in fact on the brink of instability. Gradually increasing in complexity over time to the limit which can be supported by the stability of their environment, they approach ever closer to the limits of their own stability (May, 1975d; Cohen and Newman, 1988). And as a final irony such a view of 'life at the frontier of stability' arises both from dynamical and static models of community organization; community assembly models of Robinson and Valentine (1979) and Post and Pimm (1983) suggest that as construction of the community proceeds, it becomes increasingly resistant to invasion, but actually approaches closer and closer to instability. The effect is not due to an increase in species number as in dynamical models; indeed it is unclear what does account for it (Nee, 1990). But both forms of analyses concur in a view of complexity to the limit of instability.

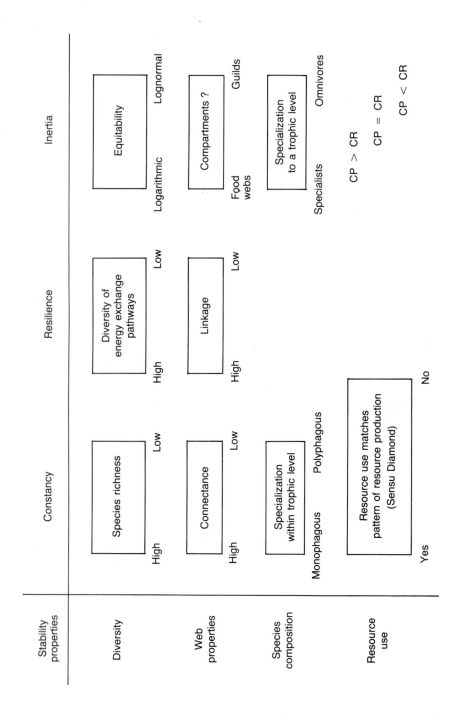

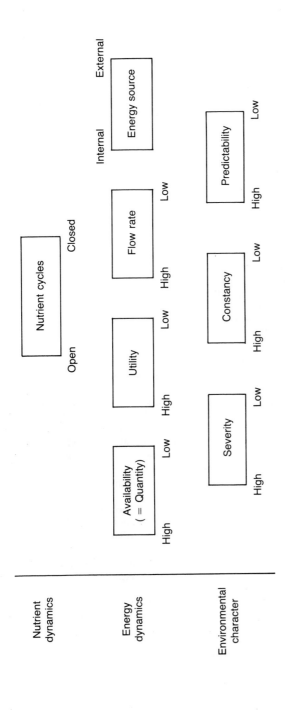

Figure 9.3 The relationship of stability properties (Constancy, Resilience and Inertia) to various characteristics of a community's structure and function. Various characteristics of community organization are shown in the figure, below a horizontal line separating these structural characteristics from derived stability properties. You are encouraged to photocopy the figure and to trace upon it the direction and nature (enhancing, decreasing) of interactions between the various community properties themselves, before tracing in connections showing the influence of such properties on the different aspects of community stability.

References

Abrams, P. A. (1984) Recruitment, lotteries and coexistence in coral reef fish. *American Naturalist*, **123**, 44–5.

Abramsky, Z. (1978) Small mammal community ecology: changes in species diversity in response to manipulated productivity. *Oecologia*, **34**, 113–123.

Adams, J. (1981) Serological analysis of the diet of *Bdellocephala punctata*, a freshwater triclad. *Oikos*, **36**, 99–106.

Adams, J. (1985) The definition and interpretation of guild structure in ecological communities. *Journal of Animal Ecology*, **54**, 43–59.

Armstrong, R. A. (1982) The effects of connectivity on community stability. *American Naturalist*, **120**, 391–402.

Arthur, W. and Mitchell, P. (1989) A revised scheme for the classification of population interactions. *Oikos*, **56**, 141–3.

Ashbourne, S. R. C. and Putman, R. J. (1987) Competition, resource-partitioning and species richness in the phytophagous insects of red oak and aspen in Canada and the U.K. *Acta Oecologia (Generalis)*, **8**, 43–56.

Atkinson, W. D. and Shorrocks, B. (1981) Competition on a divided and ephemeral resource: a simulation model. *Journal of Animal Ecology*, **50**, 461–71.

Atkinson, W. D. and Shorrocks, B. (1984) Aggregation of larval Diptera over discrete and ephemeral breeding sites: the implications for coexistence. *American Naturalist*, **124**, 336–51.

Auerbach, M. J. (1984) Stability, probability and the topology of food webs, in *Ecological Communities: Conceptual Issues and the Evidence* (eds D. R. Strong, D. Simberloff, L. G. Abele and A. B. Thistle), Princeton University Press, Princeton, pp. 413–36.

Ayala, F. (1970) Competition, coexistence and evolution, in *Essays in Evolution and Genetics* (eds M. K. Hecht and W. C. Steere), Appleton-Century-Crofts, New York.

Beauchamp, R. S. A. and Ullyett, P. (1932) Competitive relationships between certain species of freshwater triclads. *Journal of Ecology*, **20**, 200–8.

Beddington, J. R., Free, C. A. and Lawton, J. H. (1975) Dynamic complexity in predator–prey models framed in difference equations. *Nature*, **255**, 58–60.

Begon, M., Harper, J. L. and Townsend, C. R. (1986) *Ecology: Individuals, Populations and Communities*, Blackwell Scientific Publications, Oxford.

Begon, M. and Mortimer, M. (1981, 1992) *Population Ecology: A Unified Study of Animals and Plants*, Blackwell Scientific Publications, Oxford.

Bender, E. A., Case, T. J. and Gilpin, M. E. (1984) Perturbation experiments in community ecology: theory and practice. *Ecology*, **65**, 1–13.

Berryman, A. A. and Millstein, J. A, (1989) Are ecological systems chaotic – and if not, why not? *Trends in Ecology and Evolution*, **4**, 26–8.

Blondel, J., Vuilleumier, F., Marcus, L. E. and Terouanne, E. (1984) Is there ecomorphological convergence among Mediterranean bird communities of Chile, California and France? *Evolutionary Biology*, **18**, 141–213.

Botkin, D. B. (1974) Functional groups of organisms in model ecosystems, in *Ecosystem Analysis and Prediction* (ed. S. Levin), Society of Industrial and Applied Mathematics, Philadelphia, pp. 98–102.

Botkin, D. B., Janak, J. F. and Wallis, J. R. (1972) Some ecological consequences of a computer model of forest growth. *Journal of Ecology*, **60**, 849–72.

Bowers, M. A. and Brown, J. H. (1982) Body size and coexistence in desert rodents: chance or community structure? *Ecology*, **63**, 391–400.

Bradley, R. A. and Bradley, D. W. (1985) Do non-random patterns of species in niche space imply competition? *Oikos*, **45**, 443–6.

Briand, F. (1983) Environmental control of food web structure. *Ecology*, **64**, 253–63.

Briand, F. and Cohen, J. E. (1984) Community foodwebs have scale-invariant structure. *Nature*, **307**, 264–6.

Brock, T. D. and Madigan, M. T. (1988) *Biology of Microorganisms*, 5th edn, Prentice-Hall, New York.

Brown, J. H. (1973) Species diversity of seed-eating rodents in sand-dune habitats. *Ecology*, **54**, 775–87.

Brown, J. H. (1981) Two decades of homage to Santa Rosalia: toward a general theory of diversity. *American Zoologist*, **21**, 877–88.

Brown, J. H. and Davidson, D. W. (1977) Competition between seed-eating rodents and ants in desert ecosystems. *Science*, **196**, 800–2.

Brown, J. H., Davidson, D. W., Munger, J. C. and Inouye, R. S. (1986) Experimental community ecology: the desert granivore system, in *Community Ecology* (eds J. M. Diamond and T. J. Case), New York, Harper and Row, pp. 41–61.

Brown, J. H. and Kodric-Brown, A. (1977) Turnover rates in insular biogeography: the effects of immigration on extinction. *Ecology*, **58**, 445–9.

Brown, W. L. and Wilson, E. O. (1956) Character displacement. *Systematic Zoology*, **5**, 49–64.

Carothers, J. H. and Jaksic, F. M. (1984) Time as a niche difference: the role of interference competition. *Oikos*, **42**, 403–6.

Case, T. J. (1983) Niche overlap and the assembly of island lizard communities. *Oikos*, **41**, 427–433.

Chesson, P. L. and Case, T. J. (1986) Non-equilibrium community theories: chance, variability, history and coexistence, in *Community Ecology* (eds J. M. Diamond and T. J. Case), Harper and Row, New York, pp. 229–39.

Clements, F. E. (1916) Plant succession: an analysis of the development of vegetation. *Carnegie Institution of Washington Publication*, **242**, 1–512.

Clements, F. E. (1936) The nature of the climax. *Journal of Ecology*, **24**, 252–84.

Cody, M. L. (1968) On the methods of resource division in grassland bird communities. *American Naturalist*, **102**, 107–148.

Cody, M. L. (1974) *Competition and the Structure of Bird Communities*, Princeton University Press, Princeton.

Cody, M. L. (1975) Towards a theory of continental species diversities, in *Ecology and Evolution of Communities* (eds M. L. Cody and J. M. Diamond), Belknap, Harvard, pp. 214–57.

References

Cody, M. L. and Diamond, J. M. (eds) (1975) *Ecology and Evolution of Communities*, Belknap, Harvard.

Cohen, J. E. (1968) Alternative derivation of a species-abundance relation. *American Naturalist*, **102**, 165–72.

Cohen, J. E. (1978) *Food Webs and Niche Space*, Princeton University Press, Princeton.

Cohen, J. E., Briand, F. and Newman, C. M. (1986) A stochastic theory of food webs. III Predicted and observed lengths of food chains. *Proceedings of the Royal Society of London, B*, **228**, 317–53.

Cohen, J. E., Briand, F. and Newman, C. M. (1990) *Community Food Webs*, Springer-Verlag, New York.

Cohen, J. E. and Newman, C. M. (1985) A stochastic theory of community food webs. 1. Models and aggregated data. *Proceedings of the Royal Society of London, B*, **224**, 421–48.

Cohen, J. E. and Newman, C. M. (1988) Dynamic basis of food web organization. *Ecology*, **69**, 1655–64.

Cohen, J. E., Newman, C. M. and Briand, F. (1985) A stochastic theory of community food webs. II. Individual webs. *Proceedings of the Royal Society of London, B*, **224**, 449–61.

Colwell, R. K. and Futuyma, D. J. (1971) On the measurement of niche breadth and overlap. *Ecology*, **52**, 567–76.

Colwell, R. K. and Winkler, D. W. (1984) A null model for null models in biogeography, in *Ecological Communities: Conceptual Issues and the Evidence* (eds D. R. Strong, D. Simberloff, L. G. Abele and A. B. Thistle), Princeton University Press, Princeton, pp. 344–59.

Connell, J. H. (1978) Diversity in tropical rain forests and coral reefs. *Science*, **199**, 1302–10.

Connell, J. H. (1980) Diversity and the coevolution of competitors, or the ghost of competition past. *Oikos*, **35**, 131–8.

Connell, J. H. (1983) On the prevalence and relative importance of interspecific competition: evidence from field experiments. *American Naturalist*, **122**, 661–96.

Connell, J. H. and Slatyer, R. O. (1977) Mechanisms of succession in natural communities and their role in community stability and organization. *American Naturalist*, **111**, 1119–44.

Connor, E. F. and Simberloff, D. (1979) The assembly of species communities: chance or competition? *Ecology*, **60**, 1132–40.

Connor, E. F. and Simberloff, D. (1984) Neutral models of species co-occurrence patterns, in *Ecological Communities: Conceptual Issues and the Evidence* (eds D. R. Strong, D. Simberloff, L. G. Abele and A. B. Thistle), Princeton University Press, Princeton, pp. 316–31.

Cooke, G. D., Beyers, R. J. and Odum, E. P. (1969) The case for the multispecies ecological system, with special reference to succession and stability. *NASA Report SP-165*, pp. 129–39.

Coombs, C. H. (1964) *A Theory of Data*, Wiley, New York.

Cooper, W. S. (1926) The fundamentals of vegetational change. *Ecology*, **7**, 391–413.

Cowles, H. C. (1901) The physiographic ecology of Chicago and vicinity: a study of the origin, development and classification of plant societies. *Botanical Gazette*, **31**, 73–108.

Craik, J. C. A. (1989) The Gaia hypothesis – fact or fancy? *Journal of the Marine Biological Association of the UK*, **69**, 759–68.

Crawley, M. J. (1983) *Herbivory*, Blackwell Scientific Publications, Oxford.

Crawley, M. J. (1986) *Plant Ecology*, Blackwell Scientific Publications, Oxford.

Crocker, R. L. and Major, J. (1955) Soil development in relation to vegetation and surface age at Glacier Bay, Alaska. *Journal of Ecology*, **43**, 427–48.

Crowell, K. L. and Pimm, S. L. (1976) Competition and niche shifts of mice introduced onto small islands. *Oikos*, **27**, 251–8.

Davidson, J. (1938) On the growth of the sheep population in Tasmania. *Transactions of the Royal Society of South Australia*, **62**, 342–6.

DeAngelis, D. L. (1975). Stability and connectance in food web models. *Ecology*, **56**, 238–43.

DeAngelis, D. L. (1980) Energy flow, nutrient cycling and ecosystem resilience. *Ecology*, **61**, 764–71.

DeAngelis, D. L. (1992) *Dynamics of Nutrient Cycling and Food Webs*, Chapman & Hall, London.

DeAngelis, D. L., Gardner, R. H., Mankin, J. B., Post, W. U. and Carney, J. H. (1978) Energy flow and the number of trophic levels in ecological communities. *Nature*, **273**, 406–7.

Diamond, J. M. (1975) Assembly of Species Communities, in *Ecology and Evolution of Communities* (eds M. L. Cody and J. M. Diamond), Belknap, Harvard, pp. 342–444.

Diamond, J. M. and Case, T. J. (1986) *Community Ecology*, Harper & Row, New York.

Diamond, J. M. and Mayr, E. (1976) The species–area relation for birds of the Solomon archipelago. *Proceedings of the National Academy of Sciences USA*, **73**, 262–6.

Diamond, J. M. and Gilpin (1982) Examination of the 'null' model of Connor and Simberloff for species co-occurrences on islands. *Oecologia*, **52**, 64–74.

Diamond, J. M., Pimm, S. L., Gilpin, M. E. and LeCroy, M. (1989) Rapid evolution of character displacement in myzomelid honeyeaters. *American Naturalist*, **134**, 675–708.

Drake, J. A. (1990) Communities as assembled structures: do rules govern pattern? *Trends in Ecology and Evolution*, **5**, 159–64.

Drake, J. A. (1991) Community assembly mechanics and the structure of an experimental species ensemble. *American Naturalist*, **137**, 1–26.

Edwards, D. C., Conover, D. O. and Sutter, F. (1982) Mobile predators and the structure of marine intertidal communities. *Ecology*, **63**, 1175–80.

Edwards, P. J. and Wratten, S. D. (1980) *Ecology of Insect-Plant Interactions. Studies in Biology 121*, Edward Arnold, London.

Einarsen, A. S. (1945) Some factors affecting ring-necked pheasant population density. *Murrelet*, **26**, 39–44.

Elton, C. S. (1927) *Animal Ecology*, Methuen, London.

Elton, C. S. (1958) *The Ecology of Invasions by Animals and Plants*, Chapman & Hall, London.

Elton, C. S. (1966) *The Pattern of Animal Communities*, Methuen, London.

Facelli, J. and Pickett, S. T. A. (1990) Markovian chains and the role of history in succession. *Trends in Ecology and Evolution*, **5**, 27–30.

Feinsinger, P. (1976) Organisation of a tropical guild of nectarivorous birds. *Ecological Monographs*, **46**, 257–91.

Feinsinger, P. (1987) Approaches to nectarivore–plant interactions in the New World. *Review of Chilean Natural History*, **60**, 285–319.

Feinsinger, P., Swarm, L. A. and Wolf, J. A. (1985) Nectar-feeding birds on Trinidad and Tobago: comparison of diverse and depauperate guilds. *Ecological Monographs*, **55**, 1–28.

References

Fenchel, T. (1975) Character displacement and coexistence of mud snails (Hydobiidae). *Oecologia*, **20**, 19–32.

Fenchel, T. and Christiansen, F. B. (1976) *Theories of Biological Communities*, Springer-Verlag, New York.

Fenner, M. (1978a) Susceptibility to shade in seedlings of colonising and closed turf species. *New Phytologist*, **81**, 739–44.

Fenner, M. (1978b) A comparison of the abilities of colonisers and closed-turf species to establish from seed in artificial swards. *Journal of Ecology*, **66**, 953–63.

Fenner, M. (1980) The inhibition of germination of *Bidens pilosa* seeds by leaf canopy shade in some natural vegetation types. *New Phytologist*, **84**, 95–101.

Fisher, R. A., Corbet, A. S. and Williams, C. B. (1943) The relation between the number of species and the number of individuals in a random sample of an animal population. *Journal of Animal Ecology*, **12**, 42–58.

Fuentes, E. R. (1976) Ecological convergence of lizard communities in Chile and California. *Ecology*, **57**, 3–17.

Fuller, M. E. (1934) The insect inhabitants of carrion: a study in animal ecology. *CSIRA Bulletin*, **82**, 5–62.

Futuyma, D. J. and Slatkin, M. (1983) *Coevolution*, Sinauer Associates, New York.

Gardner, M. R. and Ashby, W. R. (1970) Connectance of large dynamical systems: critical values for stability. *Nature*, **288**, 784.

Gause, G. F. (1934) *The Struggle for Existence*, reprinted 1964, Hafner, New York.

Gilbert, F. S. (1980) The equilibrium theory of island biogeography: fact or fiction? *Journal of Biogeography*, **7**, 209–35.

Gilpin, M. E. and Diamond, J. M. (1984) Are species co-occurrences on islands non-random and are null hypotheses useful in community ecology? in *Ecological Communities: Conceptual Issues and the Evidence* (eds D. R. Strong, D. Simberloff, L. G. Abele and A. B. Thistle), Princeton University Press, Princeton, pp. 297–315.

Gilpin, M. E., Diamond, J. M., Connor, E. F. and Simberloff, D. (1984) Rejoinders, in *Ecological Communities: Conceptual Issues and the Evidence* (eds D. R. Strong, D. Simberloff, L. G. Abele and A. B. Thistle), Princeton University Press, Princeton, pp. 332–43.

Gladfelter, W. B. and Johnson, W. S. (1983) Feeding niche separation in a guild of tropical reef fishes. *Ecology*, **64**, 552–63.

Gladfelter, W. B., Ogden, J. C. and Gladfelter, E. H. (1980) Similarity and diversity amongst coral reef fish communities. *Ecology*, **61**, 1156–68.

Gleason, H. A. (1926) The individualistic concept of the plant association. *Bulletin of the Torrey Botanical Club*, **53**, 7–26.

Goh, B. S. (1979) Robust stability concepts for ecosystems models, in *Theoretical Systems Ecology* (ed. E. Halfon), Academic Press, London, pp. 467–87.

Goldsmith, F. B. (1973) The vegetation of exposed sea-cliffs at South Stack, Anglesey II. Experimental studies. *Journal of Ecology*, **61**, 819–29.

Grace, J. B. (1985) Juvenile vs adult competitive abilities in plants – size-dependence in cattails (*Typha*). *Ecology*, **66**, 1630–38.

Grant, P. R. (1968) Bill size, body size and the ecological adaptations of bird species to the competitive situations on islands. *Systematic Zoology*, **17**, 319–33.

Grant, P. R. and Abbott, I. A. (1980) Interspecific competition, island biogeography and null hypotheses. *Evolution*, **34**, 332–41.

Greene, E. (1987) Sizing up size ratios. *Trends in Ecology and Evolution*, **2**, 79–81.

Grossman, G. D. (1982) Dynamics and organisation of a rocky intertidal fish assemblage: the persistence and resilience of taxocene structure. *American Naturalist*, **119**, 611–37.

Gurney, W. S. C., Nisbet, R. M. and Lawton, J. H. (1983) The systematic formulation of tractable single-species population models incorporating age structure. *Journal of Animal Ecology*, **52**, 479–95.

Haefner, J. W. (1981) Avian community assembly rules: the foliage-gleaning guild. *Oecologia*, **50**, 131–42.

Haefner, P. A. (1970) The effect of low dissolved oxygen concentrations on temperature-salinity tolerance of the sand shrimp *Crangon septemspinosa*. *Physiological Zoology*, **43**, 30–37.

Haigh, J. and Maynard Smith, J. (1972) Can there be more predators than prey? *Theoretical Population Biology*, **3**, 290–9.

Hairston, N. G. (1980) The experimental test of an analysis of field distributions: competition in terrestrial salamanders. *Ecology*, **61**, 817–26.

Hairston, N. G., Smith, F. E. and Slobodkin, L. B. (1960) Community structure, population control and competition. *American Naturalist*, **44**, 421–5.

Hall, S. J. and Raffaelli, D. (1991) Food-web patterns: lessons from a species-rich web. *Journal of Animal Ecology*, **60**, 823–41.

Hanski, I. (1978) Some comments on the measurement of niche metrics. *Ecology*, **59**, 168–74.

Harper, J. L. (1975) Review of *Allelopathy* by E. L. Rice. *Quarterly Review of Biology*, **50**, 493–5.

Harper, J. L. (1977) *Population Biology of Plants*, Academic Press, London.

Harper, J. L. (1981) The concept of population in modular organisms, in *Theoretical Ecology* (ed. R. M. May), Blackwell Scientific Publications, Oxford, pp. 53–7.

Harper, J. L. and Bell, A. D. (1979) The population dynamics of growth form in organisms with modular construction, in *Population Dynamics* (eds R. M. Anderson, B. C. Turner and L. R. Taylor), Blackwell Scientific Publications, Oxford, pp. 29–52.

Harrison, G. W. (1979) Stability under environmental stress: resistance, resilience, persistence and variability. *American Naturalist*, **113**, 659–69.

Hassell, M. P. (1975) Density-dependence in single species populations. *Journal of Animal Ecology*, **44**, 283–95.

Hassell, M. P., Lawton, J. H. and May, R. M. (1976) Patterns of dynamical behaviour in single-species populations. *Journal of Animal Ecology*, **45**, 471–86.

Hastings, A. (1987) Can competition be detected using species co-occurrence data? *Ecology*, **68**, 117–23.

Hastings, A. (1988) Food web theory and stability. *Ecology*, **69**, 1665–8.

Hawkins, C. P. and MacMahon, J. A. (1989) Guilds: the multiple meanings of a concept. *Annual Review of Entomology*, **34**, 423–51.

Haydon, D. (1992) Stability and complexity revisited. Ph.D. thesis, University of Austin, Texas.

Heatwole, H. and Levins, R. (1972) Trophic structure, stability and faunal change during recolonization. *Ecology*, **53**, 531–4.

Hildrew, A., Townsend, C. R. and Hasham, A. (1985) The predatory Chironomidae of an iron-rich stream: feeding ecology and food web structure. *Ecological Entomology*, **10**, 403–13.

Hill, M. O. (1973) Diversity and evenness: a unifying notation and its consequences. *Ecology*, **54**, 427–32.

Holmes, R. T., Bonney, R. E. and Paccala, S. W. (1979) Guild structure of the Hubbard Brook bird community. *Ecology*, **60**, 512–20.

Holt, R. D. (1977) Predation, apparent competition and the structure of prey communities. *Theoretical Population Biology*, **12**, 197–229.

Holt, R. D. (1987) On the relation between niche overlap and competition: the effect of incommensurable niche dimensions. *Oikos*, **48**, 110–15.

Hope Simpson, J. F. (1940) Studies of the vegetation of the English Chalk. VI. Late stages in succession leading to chalk grassland. *Journal of Ecology*, **28**, 386–402.

Horn, H. S. (1975) Markovian Processes of Forest Succession, in *Ecology and Evolution of Communities* (eds M. L. Cody and J. M. Diamond), Belknap, Harvard, pp. 196–211.

Horn, H. S. (1976) Succession, in *Theoretical Ecology* (ed. R. M. May), Blackwell Scientific Publications, Oxford, pp. 187–204.

Horn, H. S. (1981) Succession, in *Theoretical Ecology*, 2nd edn, (ed. R. M. May), Blackwell Scientific Publications, Oxford, pp. 253–71.

Huey, R. B., Pianka, E. R., Egan, M. E. and Coons, L. W. (1974) Ecological shifts in sympatry: Kalahari fossorial lizards (*Typhlosaurus*). *Ecology*, **55**, 304–16.

Huffaker, C. B. (1958) Experimental studies on predation: dispersion factors and predator–prey oscillations. *Hilgardia*, **27**, 343–83.

Hunter, M. D. and Price, P. W. (1992) Playing chutes and ladders: heterogeneity and the relative roles of bottom-up and top-down forces in natural communities. *Ecology*, **73**, 724–32.

Hurlbert, S. H. (1978) The measurement of niche overlap and some derivatives. *Ecology*, **59**, 67–77.

Huston, M. (1979) A general hypothesis of species diversity. *American Naturalist*, **113**, 81–101.

Hutchinson, G. E. (1953) The concept of pattern in ecology. *Proceedings of the Academy of Natural Sciences, Philadelphia*, **105**, 1–12.

Hutchinson, G. E. (1957) Concluding remarks. *Cold Spring Harbor Symposia in Quantitative Biology*, **22**, 415–27.

Hutchinson, G. E. (1959) Homage to Santa Rosalia, or why are there so many kinds of animals? *American Naturalist*, **93**, 145–59.

Inger, R. F. and Colwell, R. K. (1977) Organisation of contiguous communities of amphibians and reptiles of Thailand. *Ecological Monographs*, **47**, 229–53.

Jaksic, F. M. (1981) Abuse and misuse of the term 'guild' in ecological studies. *Oikos*, **37**, 397–400.

Jaksic, F. M. and Medel, R. G. (1990) Objective recognition of guilds: testing for statistically significant species clusters. *Oecologia*, **82**, 87–92.

Jara, H. F. and Moreno, C. A. (1984) Herbivory and structure in a midlittoral rocky community – a case in southern Chile. *Ecology*, **65**, 28–38.

Jefferies, M. (1989) Measuring Talling's 'element of chance in pond populations'. *Freshwater Biology*, **21**, 383–93.

Jeffries, C. (1974) Qualitative stability and digraphs in model ecosystems. *Ecology*, **55**, 1415–9.

Joern, A. and Lawlor, L. R. (1981) Guild structure in grasshopper assemblages based on food and microhabitat resources. *Oikos*, **37**, 93–104.

Johns, A. D. (1984) Effects of selective logging on the animal communities of Malaysian hill forest. Ph.D. thesis, University of Cambridge.

Johns, A. D. (1986) Effects of selective logging on the ecological organization of a peninsular Malaysian rainforest avifauna. *Forktail*, **1**, 65–79.

Karr, J. R. (1980) Geographical variation in the avifaunas of tropical forest undergrowth. *Auk*, **97**, 283–98.

Kemeny, J. G. and Snell, J. L. (1960) *Finite Markov Chains*, Van Nostrand, New York.

Law, R. and Blackford, J. C. (1992) Self-assembling food-webs: a global viewpoint of coexistence of species in Lotka–Volterra communities. *Ecology,* **73**, 567–78.

Lawlor, L. R. (1978) A comment on randomly constructed ecosystem models. *American Naturalist,* **112**, 445–7.

Lawlor, L. R. (1980) Structure and stability in natural and randomly constructed competitive communities. *American Naturalist,* **116**, 394–408.

Lawton, J. H. (1982) Vacant niches and unsaturated communities: a comparison of bracken herbivores in two continents. *Journal of Animal Ecology,* **51**, 573–96.

Lawton, J. H. (1984) Non-competitive populations, non-convergent communities and vacant niches: the herbivores of bracken, in *Ecological Communities: Conceptual Issues and the Evidence* (eds D. R. Strong, D. Simberloff, L. G. Abele and A. B. Thistle), Princeton University Press, Princeton, pp. 67–95.

Lawton, J. H. and Strong, D. R. (1981) Community patterns and competition in folivorous insects. *American Naturalist,* **118**, 317–38.

Lawton, J. H. and Warren, P. H. (1988) Static and dynamic explanations for patterns in food webs. *Trends in Ecology and Evolution,* **3**, 242–5.

Leak, W. B. (1970) Successional change in northern hardwoods predicted by birth and death simulation. *Ecology,* **51**, 794–801.

Leuthold, W. (1978) Ecological separation among browsing ungulates in Tsavo East National Park, Kenya. *Oecologia,* **35**, 241–52.

Leslie, P. H. (1945) On the use of matrices in certain population mathematics. *Biometrika,* **33**, 183–212.

Leslie, P. H. (1948) Some further notes on the use of matrices in population mathematics. *Biometrika,* **35**, 213–45.

Levine, S. H. (1976) Competitive interactions in ecosystems. *American Naturalist,* **110**, 903–10.

Levins, S. (1968) *Evolution in Changing Environments,* Princeton University Press, Princeton.

Lewontin, R. C. (1969) The meaning of stability. *Brookhaven Symposia in Biology,* **22**, 13–24.

Lindeman, R. (1942) The tropho-dynamic aspect of ecology. *Ecology,* **23**, 399–418.

Lotka, A. J. (1925) *Elements of Physical Biology,* Williams & Wilkins, Baltimore.

MacArthur, R. H. (1955) Fluctuations of animal populations and a measure of community stability. *Ecology,* **36**, 533–6.

MacArthur, R. H. (1957) On the relative abundance of bird species. *Proceedings of the National Academy of Sciences, USA,* **43**, 293–5.

MacArthur, R. H. (1960) On the relative abundance of species. *American Naturalist,* **94**, 25–36.

MacArthur, R. H. (1968) The theory of the niche, in *Population Biology and Evolution* (ed. R. C. Lewontin), Syracuse, pp. 159–76.

MacArthur, R. H. (1970) Species packing and competitive equilibrium for many species. *Theoretical Population Biology,* **1**, 1–14.

MacArthur, R. H. (1972) *Geographical Ecology,* Harper & Row, New York.

MacArthur, R. H. and Levins, R. (1964) Competition, habitat selection and character displacement in a patchy environment. *Proceedings of the National Academy of Sciences, USA,* **51**, 1207–10.

MacArthur, R. H. and Levins, R. (1967) The limiting similarity, convergence and divergence of coexisting species. *American Naturalist,* **101**, 377–85.

MacArthur, R. H. and MacArthur, J. W. (1961) On bird species diversity. *Ecology,* **42**, 594–8.

MacArthur, R. H. and Wilson, E. O. (1967) *The Theory of Island Biogeography,* Princeton University Press, Princeton.

Magurran, A. E. (1988) *Ecological Diversity and its Measurement*, Chapman & Hall, London.

Margalef, R. and Gutierrez, E. (1983) How to introduce connectance on the frame of an expression for diversity. *American Naturalist*, **121**, 601–7.

Matson, P. A. and Hunter, M. D. (1992) The relative contributions of top-down and bottom-up forces in population and community ecology. *Ecology*, **73**, 723.

May, R. M. (1972a) Will a large complex system be stable? *Nature*, **238**, 413–4.

May, R. M. (1972b) Limit cycles in predator–prey communities. *Science*, **177**, 900–2.

May, R. M. (1973a) *Stability and Complexity in Model Ecosystems*, Princeton University Press, Princeton.

May, R. M. (1973b) Qualitative stability in model ecosystems. *Ecology*, **54**, 638–41.

May, R. M. (1975a) Some notes on measuring the competition matrix α. *Ecology*, **56**, 737–41.

May, R. M. (1975b) Patterns of Species Abundance and Diversity, in *Ecology and Evolution of Communities* (eds M. L. Cody and J. M. Diamond), Belknap, Harvard, 81–120.

May, R. M. (1975c) Biological populations obeying difference equations: stable points, stable cycles and chaos. *Journal of Theoretical Biology*, **49**, 511–24.

May, R. M. (1975d) Stability in ecosystems: some comments, in *Unifying Concepts in Ecology* (eds W. H. van Dobben and R. H. Lowe-McConnell). W. Junk, The Hague, pp. 161–8.

May, R. M. (1976, 1981) *Theoretical Ecology*, Blackwell Scientific Publications, Oxford.

May, R. M. (1984) Real and apparent patterns in community structure, in *Ecological Communities: Conceptual Issues and the Evidence* (eds D. R. Strong, D. Simberloff, L. G. Abele and A. B. Thistle), Princeton University Press, Princeton, pp. 3–16.

May, R. M. (1986a) The search for patterns in the balance of nature. *Ecology*, **67**, 1115–26.

May, R. M. (1986b) When two and two do not make four: non-linear phenomena in ecology. *Proceedings of the Royal Society of London, B*, **228**, 241–66.

May, R. M. and MacArthur, R. H. (1972) Niche overlap as a function of environmental variability. *Proceedings of the National Academy of Sciences, USA*, **69**, 1109–13.

McMurtrie, R. E. (1975) Determinants of stability of large, randomly connected systems. *Journal of Theoretical Biology*, **50**, 1–11.

McNaughton, S. J. (1978) Stability and diversity of ecological communities. *Nature*, **274**, 251–3.

McNaughton, S. J., Oesterheid, M., Frank, D. A. and Williams, K. J. (1989) Ecosystem – level patterns of primary productivity and herbivory in terrestrial habitats. *Nature*, **341**, 142–4.

Menge, B. A. (1972) Competition for food between two intertidal starfish species, and its effect on bodysize and feeding. *Ecology*, **53**, 635–44.

Menge, B. A. (1976) Organization of the New England rocky intertidal community: role of predation, competition and environmental heterogeneity. *Ecological Monographs*, **46**, 355–93.

Menge, B. A. (1992) Community regulation: under what conditions are bottom-up factors important on rocky shores? *Ecology*, **73**, 755–65.

Menge, B. A. and Sutherland, J. P. (1976) Species diversity gradients: synthesis of the roles of predation, competition and temporal heterogeneity. *American Naturalist*, **110**, 351–69.

Miller, T. E. (1982) Community diversity and interactions between the size and frequency of disturbance. *American Naturalist*, **120**, 533–6.

Motomura, I. (1932) A statistical treatment of associations. *Japanese Journal of Zoology*, **44**, 279–383.

Moulton, M. P. (1985) Morphological similarity and coexistence of congeners: an experimental test with introduced Hawaiian birds. *Oikos*, **44**, 301–5.

Moulton, M. P. and Pimm, S. L. (1983) The introduced Hawaiian avifauna: biogeographic evidence for competition. *American Naturalist*, **121**, 669–90.

Moulton, M. P. and Pimm, S. L. (1986) The extent of competition in shaping an introduced avifauna, in *Community Ecology* (eds J. M. Diamond and T. J. Case), Harper & Row, New York, pp. 80–97.

Nee, S. (1990) Community construction. *Trends in Ecology and Evolution*, **5**, 337–40.

Nicholson, A. J. and Bailey, V. A. (1935) The balance of animal populations. *Proceedings of the Zoological Society of London*, **3**, 551–98.

Nilsson, L. (1969) Food consumption of diving ducks wintering at the coast of South Sweden. *Oikos*, **20**, 128–35.

Odum, E. P. (1959/1971) *Fundamentals of Ecology*, W. B. Saunders, New York.

Odum, E. P. (1969) The strategy of ecosystem development. *Science*, **164**, 262–70.

Odum, E. P. (1975) Diversity as a function of energy flow, in *Unifying Concepts of Ecology* (eds W. H. van Dobben and R. H. Lowe-McConnell), W. Junk, The Hague, pp. 11–14.

Odum, H. T. (1957) Trophic structure and productivity of Silver Springs, Florida. *Ecological Monographs*, **27**, 55–112.

Oksanen, L. (1987) Interspecific competition and the structure of bird guilds in boreal Europe: the importance of doing fieldwork in the right season. *Trends in Ecology and Evolution*, **2**, 376–9.

Orians, G. H. (1975) Diversity, stability and maturity in natural ecosystems, in *Unifying Concepts in Ecology* (eds W. H. van Dobben and R. H. Lowe-McConnell), W. Junk, The Hague, pp. 139–50.

Orians, G. H. and Horn, H. S. (1969) Overlap in foods and foraging of four species of blackbirds in the potholes of central Washington. *Ecology*, **50**, 930–8.

Osman, R. W. (1977) The establishment and development of a marine epifaunal community. *Ecological Monographs*, **47**, 37–63.

Paine, R. T. (1966) Food web complexity and species diversity. *American Naturalist*, **100**, 65–75.

Paine, R. T. (1969) A note on trophic complexity and community stability. *American Naturalist*, **103**, 91–3.

Paine, R. T. (1980) Foodwebs: linkage, interaction strength and community infrastructure. *Journal of Animal Ecology*, **49**, 667–86.

Paine, R. T. (1988) Foodwebs: road maps of interactions, or grist for theoretical development? *Ecology*, **69**, 1648–54.

Park, T. (1954) Experimental studies of interspecies competition. II. Temperature, humidity and competition in two species of *Tribolium*. *Physiological Zoology*, **27**, 177–238.

Park, T., Leslie, P. H. and Mertz, D. B. (1964) Genetic strains and competition in populations of *Tribolium*. *Physiological Zoology*, **37**, 97–162.

Patrick, R. (1963) The structures of diatom communities under varying ecological conditions. *Annals of the New York Academy of Science*, **108**, 353–8.

Pearl, R. (1925) *The Biology of Population Growth*, Knopf, New York.

Pianka, E. R. (1967) On lizard species diversity: North American flatland deserts. *Ecology*, **48**, 333–51.

Pianka, E. R. (1973) The structure of lizard communities. *Annual Review of Ecology and Systematics*, **4**, 53–74.
Pianka, E. R. (1975) Niche relations of desert lizards, in *Ecology and Evolution of Communities* (eds M. L. Cody and J. M. Diamond), Belknap, Harvard, pp. 291–314.
Pianka, E. R. (1976) Competition and Niche Theory, in *Theoretical Ecology* (ed. R. M. May), Blackwell Scientific Publications, Oxford, pp. 114–41.
Pianka, E. R. (1980) Guild structure in desert lizards. *Oikos*, **35**, 194–201.
Pianka, E. R. (1981) Competition and Niche Theory, in *Theoretical Ecology* (2nd edn) (ed. R. M. May), Blackwell Scientific Publications, Oxford, pp. 167–96.
Pianka, E. R., Huey, R. B. and Lawlor, L. R. (1979) Niche segregation in desert lizards, in *Analysis of Ecological Systems* (eds D. J. Horn, R. Mitchell and G. R. Stairs), Ohio State University Press, pp. 67–115.
Pielou, E. C. (1969) *Ecological Diversity*, Wiley Interscience, New York.
Pimentel, D. (1961) Species diversity and insect population outbreaks. *Annals of the Entomological Society of America*, **54**, 76–86.
Pimm, S. L. (1979) Complexity and stability; another look at MacArthur's original hypothesis. *Oikos*, **33**, 351–7.
Pimm, S. L. (1980a) Properties of food webs. *Ecology*, **61**, 219–25.
Pimm, S. L. (1980b) Bounds on foodweb connectance. *Nature*, **285**, 591.
Pimm, S. L. (1982) *Food Webs*, Chapman & Hall, London.
Pimm, S. L. (1984) The complexity and stability of ecosystems. *Nature*, **307**, 321–6.
Pimm, S. L. (1991) *The Balance of Nature*, Chicago University Press, Chicago.
Pimm, S. L. and Kitching, R. L. (1987) The determinants of foodchain lengths. *Oikos*, **50**, 302–7.
Pimm, S. L. and Lawton, J. H. (1977) The number of trophic levels in ecological communities. *Nature*, **268**, 329–31.
Pimm, S. L. and Lawton, J. H. (1978) On feeding on more than one trophic level. *Nature*, **275**, 542–4.
Pimm, S. L. and Lawton, J. H. (1980). Are food webs divided into compartments? *Journal of Animal Ecology*, **49**, 879–98.
Pimm, S. L., Lawton, J. H. and Cohen, J. E. (1991) Food web patterns and their consequences. *Nature*, **350**, 669–74.
Poole, R. W. (1974) *An Introduction to Quantitative Ecology*, McGraw-Hill, New York.
Post, W. M. and Pimm, S. L. (1983) Community assembly and food web stability. *Mathematical Bioscience*, **64**, 169–92.
Power, M. E. (1992) Top-down and bottom-up forces in food webs: do plants have primacy? *Ecology*, **73**, 733–46.
Preston, F. W. (1948) The commonness and rarity of species. *Ecology*, **29**, 254–83.
Preston, F. W. (1962) The canonical distribution of commonness and rarity. *Ecology*, **43**, 185–215; 410–432.
Pulliam, H. R. (1975) Coexistence of sparrows: a test of community theory. *Science*, **189**, 474–6.
Putman, R. J. (1984) The Geography of Animal Communities, in *Themes in Biogeography* (ed. J. A. Taylor), Croom Helm, Beckenham, pp. 163–90.
Putman, R. J. (1986) Competition and coexistence in a multispecies grazing system. *Acta Theriologica*, **31**, 271–91.
Putman, R. J. and Wratten, S. D. (1984). *Principles of Ecology*, Chapman & Hall, London.
Raffaelli, D. and Hall, S. J. (1992) Compartments and predation in an estuarine food web. *Journal of Animal Ecology*, **61**, 551–60.

Rappoldt, C. and Hogeweg, P. (1980) Niche packing and number of species. *American Naturalist*, **116**, 480–92.

Rejmanek, M. and Stary, P. (1979) Connectance in real biotic commmunities and critical values for stability of model ecosystems. *Nature*, **280**, 311–3.

Reynoldson, T. B. and Bellamy, L. S. (1970) The establishment of interspecific competition in field populations, with an example of competition in action between *Polycelis nigra* (Mull.) and *Polycelis tenuis* (Ijima), in *Dynamics of Populations* (eds P. J. den Boer and G. R. Gradwell), Pudoc, Wageningen, pp. 282–94.

Reynoldson, T. B. and Davies, R. W. (1970) Food niche and coexistence in lake-dwelling triclads. *Journal of Animal Ecology*, **39**, 599–617.

Reynoldson, T. B. and Sefton, A. D. (1976) The food of *Planaria torva* (Muller) (Turbellaria: Tricladida): a laboratory and field study. *Freshwater Biology*, **6**, 383–93.

Ricklefs, R. E. and Travis, J. (1980) A morphological approach to the study of avian community organisation. *Auk*, **97**, 321–38.

Roberts, F. S. (1971) Signed graphs and the growing demand for energy. *Environmental Planning*, **3**, 395–410.

Robinson, J. V. and Dickerson, J. E. (1984) Testing the invulnerability of laboratory island communities to invasion. *Oecologia*, **61**, 169–74.

Robinson, J. V. and Dickerson, J. E. (1987) Does invasion sequence affect community structure? *Ecology*, **68**, 587–95.

Robinson, J. V. and Edgemon, M. A. (1988) An experimental evaluation of the effect of invasion history on community structure. *Ecology*, **69**, 1410–17.

Robinson, J. V. and Valentine, W. D. (1979) The concepts of elasticity, invulnerability and invadability. *Journal of Theoretical Biology*, **81**, 91–104.

Roff, D. A. (1974) Spatial heterogeneity and the persistence of populations. *Oecologia*, **15**, 245–58.

Root, R. B. (1967) The niche exploitation pattern of the blue-grey gnat-catcher. *Ecological Monographs*, **37**, 317–50.

Roughgarden, J. (1974) Species packing and the competition function, with illustrations from coral reef fish. *Theoretical Population Biology*, **5**, 163–86.

Roughgarden, J. (1986) A comparison of food-limited and space-limited animal competition communities, in *Community Ecology* (eds J. M. Diamond and T. J. Case), Harper & Row, New York, pp. 492–516.

Roughgarden, J. and Diamond, J. (1986) The role of species interactions in community ecology, in *Community Ecology* (eds J. M. Diamond and T. J. Case), Harper & Row, New York, pp. 333–43.

Roughgarden, J. and Feldman, M. (1975) Species packing and predation pressure. *Ecology*, **56**, 489–92.

Rummel, J. D. and Roughgarden, J. (1983) Some differences between invasion-structured and coevolution-structured competitive communities: a preliminary theoretical analysis. *Oikos*, **41**, 477–86.

Rusterholz, K. A. (1981) Competition and the structure of an avian foraging guild. *American Naturalist*, **118**, 173–90.

Ryti, R. T. and Gilpin, M. E. (1987) The comparative analysis of species occurrence patterns in archipelagoes. *Oecologia*, **73**, 282–7.

Sale, P. F. (1977) Maintenance of high diversity in coral reef fish communities. *American Naturalist*, **111**, 337–59.

Sale, P. F. (1979) Recruitment, loss and coexistence in a guild of territorial coral reef fishes. *Oecologia*, **42**, 159–77.

Sale, P. F. and Williams, D. M. (1982) Community structure of coral-reef fishes –

are the patterns more than those expected by chance? *American Naturalist*, **120**, 121–7.

Sanders, H. (1968) Marine benthic diversity: a comparative study. *American Naturalist*, **102**, 243–82.

Schaffer, W. M. (1984) Stretching and folding in lynx fur returns – evidence for a strange attractor in nature. *American Naturalist*, **124**, 798–820.

Schaffer, W. M. (1987), in *Evolution of Life Histories: Theory and Patterns from Mammals* (ed. M. Boyce), Yale University Press, Yale.

Schaffer, W. M. and Kot, M. (1985a) Do strange attractors govern ecological systems? *BioScience*, **35**, 342–50.

Schaffer, W. M. and Kot, M. (1985b) Nearly one-dimensional dynamics in an epidemic. *Journal of Theoretical Biology*, **112**, 403–27.

Schaffer, W. M. and Kot, M. (1986) Chaos in ecological systems: the coals that Newcastle forgot. *Trends in Ecology and Evolution*, **3**, 58–63.

Schluter, D. (1986) Tests for similarity and convergence of finch communities. *Ecology*, **67**, 1073–85.

Schoener, T. W. (1965) The evolution of bill size differences among sympatric congeneric species of birds. *Evolution*, **19**, 189–213.

Schoener, T. W. (1974) Resource partitioning in ecological communities. *Science*, **185**, 27–39.

Schoener, T. W. (1983) Field experiments on interspecific competition. *American Naturalist*, **122**, 240–85.

Shelford, V. E. (1913) The reactions of certain animals to gradients of evaporating power and air. A study in experimental ecology. *Biological Bulletin*, **25**, 79–120.

Silvertown, J. W. (1982/1987) *Introduction to Plant Population Ecology*, Longman, Harlow.

Silvertown, J. and Law, R. (1987) Do plants need niches? Some recent developments in plant community ecology. *Trends in Ecology and Evolution*, **2**, 24–6.

Simberloff, D. S. (1976) Trophic structure determination and equilibrium in an arthropod community. *Ecology*, **57**, 395–8.

Simberloff, D. S. (1978a) Using island biogeographic distributions to determine if colonisation is stochastic. *American Naturalist*, **112**, 713–26.

Simberloff, D. S. (1978b) in *Diversity of Insect Faunas, Symposia of the Royal Entomological Society of London*, **9**, 139–53.

Simberloff, D. S. and Wilson, E. O. (1969) Experimental zoogeography of islands: the colonisation of empty islands. *Ecology*, **50**, 278–96.

Slobodkichoff, C. N. and Schulz, W. C. (1980) Measures of niche overlap. *Ecology*, **61**, 1051–5.

Slobodkin, L. B. (1961) *Growth and Regulation of Animal Populations*, Holt, Rinehart & Winston, New York.

Slobodkin, L. B. and Sanders, H. L. (1969) On the contribution of environmental predictability to species diversity. *Brookhaven Symposia in Biology*, **22**, 82–95.

Smith, E. P. and Zaret, T. M. (1982) Bias in estimating niche overlap. *Ecology*, **63**, 1248–53.

Sousa, W. P. (1979) Disturbance in marine intertidal boulder fields: the non-equilibrium maintenance of species diveristy. *Ecology*, **60**, 1225–39.

Southwood, T. R. E. (1978) *Ecological Methods*, Blackwell Scientific Publications, Oxford.

Southwood, T. R. E., Brown, V. K. and Reader, P. M. (1979) The relationships of plant and insect diversities in succession. *Biological Journal of the Linnean Society*, **12**, 327–48.

Spiller, D. A. (1984) Seasonal reversal of competitive advantage between two spider species. *Oecologia*, **64**, 322–31.

Stenseth, N. C. (1979) Where have all the species gone? On the nature of extinction and the Red Queen hypothesis. *Oikos*, **33**, 196–227.

Strong, D. R. (1992) Are trophic cascades all wet? Differentiation and donor control in speciose ecosystems. *Ecology*, **73**, 747–54.

Strong, D. R., Lawton, J. H. and Southwood, T. R. E. (1984) *Insects on Plants: Community Patterns and Mechanisms*, Blackwell Scientific Publications, Oxford.

Strong, D. R., Simberloff, D., Abele, L. G. and Thistle, A. B. (eds) (1984) *Ecological Communities: Conceptual Issues and the Evidence*, Princeton University Press, Princeton.

Sugihara, G. (1980) Minimal Community structure: an explanation of species-abundance patterns. *American Naturalist*, **116**, 770–87.

Sugihara, G. C. (1984) Graph theory, homology and food webs. *Symposia in Applied Mathematics* (Mathematical Society of America), **30**, 83–101.

Sugihara, G. and May, R. M. (1990) Non-linear forecasting as a way of distinguishing chaos from measurement error in time series. *Nature*, **344**, 734–41.

Sugihara, G., Grenfell, B. and May, R. M. (1990) Distinguishing error from chaos in ecological time series. *Philosophical Transactions of the Royal Society*, B, **330**, 235–49.

Sugihara, G., Schoenly, K. and Trombla, A. (1989) Scale-invariance on food-web properties. *Science*, **245**, 48–52.

Takens, F. (1981) Detecting strange attractors in turbulence, in *Dynamical Systems and Turbulence* (eds D. A. Rand and L-S Young), Springer-Verlag, New York, pp. 366–81.

Talling, J. F. (1951) The element of chance in pond populations. *The Naturalist (October/December 1951)*, pp. 157–70.

Tansley, A. R. (1922) Studies of the vegetation of the English Chalk. II. Early stages of redevelopment of woody vegetation in chalk grassland. *Journal of Ecology*, **10**, 168–77.

Tansley, A. R. and Adamson, R. W. (1925) Studies of the vegetation of the English Chalk. III. The chalk grasslands of the Hampshire-Sussex border. *Journal of Ecology*, **13**, 177–223.

Terborgh, J. and Robinson, S. (1986) Guilds and their utility in Ecology, in *Community Ecology: Pattern and Process* (eds J. Kikkawa and D. J. Anderson), Blackwell Scientific Publications, Oxford, pp. 65–90.

Thorman, S. (1982) Niche dynamics and resource partitioning in a fish guild inhabiting a shallow estuary on the Swedish west coast. *Oikos*, **39**, 32–9.

Tilman, D. (1980) Resources: a graphical–mechanistic approach to competition and predation. *American Naturalist*, **116**, 362–93.

Tilman, D. (1982) *Resource Competition and Community Structure*, Princeton University Press, Princeton.

Timan, D. (1986) Resources, competition and the dynamics of plant communities, in *Plant Ecology* (ed. M. J. Crawley), Blackwell Scientific Publications, Oxford, pp. 51–75.

Ugland, K. I. and Gray, J. S. (1982) Log normal distributions and the concept of community equilibrium. *Oikos*, **39**, 171–8.

Usher, M. B. (1979) Markovian approaches to ecological succession. *Journal of Animal Ecology*, **48**, 413–26.

Van Valkenburgh, B. (1988) Trophic diversity in past and present guilds of large predatory mammals. *Paleobiology*, **14**, 155–73.

Vandermeer, J. H. (1972) Niche theory. *Annual Review of Ecology and Systematics,* **3,** 107–32.

Vandermeer, J. H. (1980) Indirect mutualism: variations on a theme by Stephen Levine. *American Naturalist,* **116,** 441–8.

Verhulst, P. F. (1838) Notice sur la loi que la population suit dans sa accroisement. *Correspondence Mathematique et Physique,* **10,** 113–21.

Viereck, L. A. (1966) Plant succession and soil development on gravel outwash of the Muldrow Glacier, Alaska. *Ecological Monographs,* **36,** 181–99.

Volterra, V. (1926) Variations and fluctuations of the number of individuals in animal species living together, reprinted in Chapman, R. N. (1931) *Animal Ecology,* McGraw-Hill, New York.

Waggoners, P. E. and Stephens, G. R. (1971) Transition probabilities for a forest. *Nature,* **255,** 1160–1.

Warner, R. R. and Chesson, P. L. (1985) Coexistence mediated by recruitment fluctuations: a field guide to the storage effect. *American Naturalist,* **125,** 769–87.

Warren, P. H. (1990) Variation in food-web structure: the determinants of connectance. *American Naturalist,* **136,** 689–700.

Warren, P. H. and Lawton, J. H. (1987) Invertebrate predator–prey bodysize relationships: an explanation for upper triangular food webs and patterns in food web structure? *Oecologia,* **74,** 231–5.

Watkinson, A. R. (1980) Density-dependence in single species populations of plants. *Journal of Theoretical Biology,* **83,** 345–57.

Watkinson, A. R. (1981) Interference in pure and mixed populations of *Agrostemma githago* L. *Journal of Applied Ecology,* **18,** 967–76.

Watkinson, A. R. (1986) Plant population dynamics, in *Plant Ecology* (ed. M. J. Crawley), Blackwell Scientific Publications, Oxford, pp. 137–84.

Werner, P. A. and Caswell, H. (1977) Population growth rates and age-versus stage-distribution models for teasel (*Dipsacus sylvestris* Huds.) *Ecology,* **58,** 1103–11.

Whittaker, R. H. (1967) Gradient analysis of vegetation. *Biological Reviews,* **42,** 207–64.

Whittaker, R. H. (1975) *Communities and Ecosystems,* Macmillan, London and New York.

Wiens, J. A. (1977) On competition and variable environments. *American Scientist,* **65,** 590–7.

Wiens, J. A. (1986) Spatial scale and temporal variation in studies of shrub-steppe birds, in *Community Ecology* (eds J. M. Diamond and T. J. Case), Harper & Row, New York, pp. 154–72.

Wiens, J. A., Addicott, J. F., Case, T. J. and Diamond, J. M. (1986) The importance of spatial and temporal scale in ecological investigations, in *Community Ecology* (eds J. M. Diamond and T. J. Case), Harper and Row, New York, pp. 145–53.

Wiens, J. A. and Rotenberry, J. T. (1980) Bird community structure in cold shrub deserts: competition or chaos? *Proceedings of the XVII International Ornithological Congress, Berlin,* 1063–70.

Williams, C. B. (1953) The relative abundance of different species in a wild animal population. *Journal of Animal Ecology,* **22,** 14–31.

Williamson, M. H. (1981) *Island Ecology,* Oxford University Press, Oxford.

Winemiller, K. O. (1989) Must connectance decline with species richness? *American Naturalist,* **134,** 960–8.

Wright, S. J. and Biehl, C. C. (1982) Island biogeographic distributions: testing for random, regular and aggregated patterns of species occurrence. *American Naturalist,* **119,** 345–57.

Yodzis, P. (1980) The connectance of real ecosystems. *Nature*, **284**, 544–5.
Yount, J. L. (1956) Factors that control species numbers in Silver Springs, Florida. *Limnology and Oceanography*, **1**, 286–95.

Index

Page references in **bold** represent figures, those in *italics* represent boxes.

Algal assemblies 37–8, 107, 122–3
Allelopathy 25
American beech 99
Anatopynia pennipes 44
Asellus 67
Aspen 82
Assembly rules for communities 100–5
 assembly, chance or competition 102–8

Bdellocephala punctata 86–7
Blackgum 99
Body size distribution patterns 3, 65, 103–5
Bracken 81–2, 119
 herbivores 81–2, 119

Canadian lynx 114
Capreolus capreolus 69–70, 75
Carcinus minas 48
Cattle 69–70, 77
Cervus nippon 69–70
Chaos 112–14
Character displacement 64–5
Colonization and extinction in determining species composition of communities 120–2
Community assembly 93, 100–8
 the role of history 106–8
Compartments in food webs 52
 compartmentation and stability 150
Competition
 competitive coexistence 30–3, 109, 122–3

competitive exclusion 29–30
competitive mutualism 38
competitive release 38
diffuse, indirect mutualism and other indirect interactions 37–9, 141
effects of physical heterogeneity 30–31
effects of temporal heterogeneity 31–3, 122–3
in natural communities 33–7
interference or exploitation **25**, 25–7, 76–8
models 17–25
Competitive interactions, *see* Competition
Connectance
 and community stability 45, 146–8
 and environmental stability 46, 51–2, 146–8
 and species richness 44–52
 see also Food webs
Constancy 136–7, 145–9, 153–5, **156–7**
Coral reef fish 33, 80
Crangon septemspinosa 92
Cyclosa turbinata 32

Dama dama 69–70, 75
Dendrocoelum lacteum 67, 86
Diceros bicornis 69
Dipodomys ordii 117
Diseases 114
Dispersal 91–3
 see also Colonization

176 Index

Diversity, see Species diversity
Donor control 42, 141, 148
Drosophila pseudoobscura **25**, 31
Drosophila serrata **25**, 31
Dugesia polychroa 67, 86

Ecological communities, definitions 2
Ecological niche, see Niche
Ecological succession 94–100
　Markovian models 98–100
　mechanisms 95–8, 150–2
Energy flow within communities 8–11, 152–3
　efficiency of energy transfer between trophic levels 10–11
　implications for stability 152–3
Environmental predictability
　and niche dynamics 64
　and species diversity 124–5, 132–3, 146
　and stability 146, 153
Environmental heterogeneity
　and population interactions 30–3
　and species diversity 118
Environmental severity
　and species diversity 123–5, 146
Eotetranychus sexmaculatus 31
Equilibrium and non-equilibrium communities 109–12
Equitability 115
Eupomacentrus apicalis 33

Fallow deer 69–70, 75
Flatworms 35, 67, 86
Food refuge 67
Food webs 13, 40–59, 147–50
　compartmentation and stability 150
　connectance 44–52
　　and stability 146–8
　essential assumptions 41–2
　interaction strength 45
　　and stability 148–9
　linkage density 4, 46, 47, 148
　upper triangularity 55–7
　web structure and stability 41, 44–6, 52–7, 145–9
　web topology 52–5
　　and stability 53–5
Food web patterns, static or dynamic explanations 57–9

Gerenuk 69
Giraffa camelopardalis 69
Giraffe, see *Giraffa camelopardalis*
Grasshopper assemblies 85, **85**
Grey birch 99
Gross ecological efficiency 10
Guilds 80–8, 102, 150
　definitions 83
　guild structure 83–7

Herbivory 21, 29
Hypacanthus 53, **53**, 55

Inertia 136, 147–9, **156–7**
Interaction strength
　and stability 148–9
　see also Food webs
Invasion history, effects on community composition 106–8
Island biogeography theory 91, 120–1

Jeffries colour test 142–5
Johnius 53, **53**, 55

Kudu 69

Laboratory microcosms 44, 107
Large herbivore assemblies 69–70
Lechriodus fletcheri 44
Lepasterias hexactis 37
Limits to number of trophic levels 4, 7, 42–4
Lithognathus 53, **53**, 55
Litocranius walleri 69
Littorina sp. 48
Lizard assemblies 65, 73, 80, 83, 103, 118
Local stability 137, 146, 147
Lotka–Volterra population models 15–18, 41, 148
Lottery model of competition 32–3, 109

Mangrove communities 7–8, 80–1, 138–9
Metepeira grinnelli 32
Myrmotherula spp. 66, **66**
Mytilis edulis 37

Index

Niche
  breadth 61, 63–4
  definitions 60
  dynamics, implications of resource availability and predictability 64
  fundamental 61–2
  overlap 66–72
    and competition 67, 74–9
    limits to 71–2
    problems of calculation and interpretation 74–9
  packing and community structure 13, 72–4
  parallel 80
  realized 62–4
  separation 63, 64–6
  shifts 64–5
Niche theory
  and resource partitioning 61–74
  and implications for community structure 72–4, 75–9
Non-equilibrium community structure 110–12

Omnivory 41, 42

Paramecium bursaria 30, **30**, 107
Paramecium caudatum 30, **30**
Pisaster ochraceus 37
Planaria gonocephala 63
Planaria montenegrina 63
Plectroglyphidodon lachrymatus 33
Polycelis nigra 35, 67, 86
Polycelis tenuis 35, 67, 86–7
Polycelis torva 86–7
Pomacentrus wardi 33
Ponies 69–70
Population interactions
  dynamics of interacting species
    competition 17, *19–20*, *21*, 22–5
    predation 17, *19–20*
  effects of physical or biotic heterogeneity 30–3, 141
  implications for community structure and stability 28–33, 138, 141
Population models
  for interacting species 17–21, 22–2
  for plant populations 20, *21*
  for single species 15–16, *19–20*, *21*

Populus tremula 82
Populus tremuloides 82
Predator–prey interactions
  effects of temporal or spatial heterogeneity 30, 141
Primary productivity 11, 12, 117, 151–2
Priority effects in community assembly 106–8
Pteridium aquilinum 81–2, 119
Ptethodon glutinosus 36
Ptethodon jordani 36

Qualitative stability 142–5
Quercus rubra 82

Red oak 82
Red maple 99
Resilience 136–7, 140, 145–9
Resource partitioning
  in different communities, parallels in 80–3
  overdispersion in resource space 3, 63, 65–6
  see also Niche separation
Rhabdosargus 53, **53**, 55
Rhinoceros 69
Roe deer 69–70, 75

Salamanders 36
Sandshrimp 92
Saturation of community structure 119
Shelford's laws of tolerance 90–1
Shrub bird assemblies 73, 80, 110
Sika deer 69–70, 77
Size ratios 65, 103–5
Small rodent assemblies 114, 117
Species abundance curves 126–34
  logarithmic 127–8, 131, 132–3
  lognormal 127–8, 131, 132–3
Species diversity 115–34
  definitions 115
  effects of intermediate levels of disturbance 109, 122–3
  and environmental heterogeneity 118
  and productivity 117
  and stability 132–3, 145–7
  and structural complexity 118
Species richness
  and connectance 44–52

178 Index

Species richness (*cont'd*)
  factors determining 116–26
  see also Species diversity
Stability 135–57
  of communities, evidence for constancy, resilience and inertia 138–40
  definitions 136–7
  of populations 27–9, 141–5
  qualitative 142–5
Succession
  and stability 150–52
  see also Ecological succession

Talling's element of chance 108

Tautoglobarus 48
Thrips imaginis 114
Tilman's models of competition 22–5
Tolerance limits to abiotic conditions 61, 90–1
Tragelaphus strepsiceros 69
Tree hole communities 44
Tribolium castaneum 31
Tribolium confusum 31
Trophodynamic analyses of community structure 5–12
Typhlodromus occidentalis 31
Typhlosaurus gariepensis 65
Typhlosaurus lineatus 65